PROBABLE TOMORROWS

Probable Tomorrows

◆

HOW SCIENCE

AND TECHNOLOGY

WILL TRANSFORM

OUR LIVES

IN

THE NEXT

TWENTY YEARS

◆

MARVIN CETRON
AND
OWEN DAVIES

St. Martin's Press
New York

PROBABLE TOMORROWS: HOW SCIENCE AND TECHNOLOGY
WILL TRANSFORM OUR LIVES IN THE NEXT TWENTY YEARS.
Copyright © 1997 by Marvin Cetron and Owen Davies.
All rights reserved.
Printed in the United States of America.
No part of this book may be used or reproduced in any manner whatsoever
without written permission except in the case of brief quotations
embodied in critical articles or reviews. For information,
address St. Martin's Press, 175 Fifth Avenue,
New York, N.Y. 10010

Design by Jenny Dossin

Library of Congress Cataloging-in-Publication Data
Cetron, Marvin J.
Probable tomorrows : how science and technology will transform our lives
in the next twenty years / Marvin Cetron and Owen Davies.
p.   cm.
ISBN 0-312-15429-1
1. Science—United States—Forecasting.
2. Technology—United States—Forecasting.
3. Science—Social aspects—United States.
4. Technology—Social aspects—United States.
I. Davies, Owen L.
II. Title.
Q127.U6C28   1997
303.48'3'0112—dc21   96-54503
CIP

First Edition: June 1997
10   9   8   7   6   5   4   3   2   1

*For Kimberley—
the newest member
of our family.*

*Wishing her many
happy tomorrows
with Adam.*

CONTENTS

This book is about the future of technology. In it we will examine some of the many recent developments in a few key fields and try, in a limited way, to forecast where they will take us in the next fifteen years or so.

If that sounds like a modest goal, it's not. Technology is the dominant force of our time and probably of all time to come. It appears in more varieties than we can count. It changes so rapidly that no scientist or engineer can keep up with his own field, much less with technology in general. It permeates and shapes our lives at every turn. We live in technology as fish live in the sea, and we have only a little better chance of forecasting the details of its changes.

Yet the task is well worth undertaking. Whatever hints we can glean about the future will help us prepare for the changes to come. Modest forecasts, evidence of trends, a few concrete developments to be expected all are better than no warning at all. And though technology has made the present much less stable than the past, and surely will make the future more turbulent still, there is good reason to hope that our lives, in sum and on average, will be better as a result. In an age of uncomfortable challenges, this is reassurance we all can use.

For an idea of what is to come—in magnitude if not in specifics—look to the past. In the last ninety years, the world has shrunk, while human experience has expanded almost beyond the recognition of those who grew up in our grandparents' generation.

A century after America's founders conceived their agrarian democracy, nearly all their descendents still lived on small farms. Since World War I, technology has extracted us from behind horse-drawn plows and plugged us into assembly lines and offices. Today it is removing many of us from offices and letting us work at home or compelling us to work on the road.

As recently as 1920, the average American baby could expect to live only fifty-four years. By the early 1990s, average life expectancy in the United States had climbed to seventy-five years, seventy-two for men and nearly seventy-nine for women. In the next twenty years, life expectancy may well rise again, even more steeply. This time it will climb, not only for the newborn but for those already well into adulthood.

In transportation and communications, the changes have been even more pronounced. As recently as World War II, the average American lived and died within 38 miles (61 kilometers) of his birthplace. For New Yorkers, the radius was only 17.5 miles (28 kilometers), as far as the subway ran. Information from the outside came by newspaper, radio, or word from the traveler's mouth; it moved intermittently and often arrived only after long delay. In 1945, when the first atomic bomb fused the sand of Alamogordo, New Mexico, the shot was not heard 'round the world; rumors of a massive explosion in the desert were easily contained. Only a half century later, someone born in Massachusetts is more likely than not to attend college in Chicago, find a job in Seattle, vacation in Mexico, and retire in Florida. News from London, Moscow, Sarajevo, and Pyongyang arrives instantly on CNN and, for growing numbers of people, on personal computers fed by the Internet. From our offices in suburban Virginia and rural New Hampshire, Paris, Singapore, Buenos Aires, and Sydney are all as close as Washington and Boston; none is more distant than the few steps to the computer. Around the globe, we will spend the rest of our lives finding things to say to people we will never meet in person. Thus far, shared interests have proved easy to find.

We do not wish to make more of this than technology itself has to offer. In simpler times, early futurists occasionally forecast that science would bring world peace—often by making war too terrible to contemplate—or universal prosperity. We will not make the same mistake.

In the 1950s and 1960s, the so-called Green Revolution brought hybrid crops and new agricultural methods that multiplied farm yields. Even today, with a far larger global population, the world produces more food than it really needs. Yet there is recurring famine in Africa, less severe of late than in some years, but essentially unchanged from any period in recent memory. Technology has yet to overcome the limitations of Third World transport and tribal politics.

Since the 1940s and 1950s, antibiotics have all but eliminated infectious disease from developed lands. In all the world, smallpox exists only in the refrigerators of two closely guarded research laboratories. Yet cholera, diphtheria, and even the flu kill tens of thousands of people each year. In the United States, polio and tuberculosis are making a comeback. And AIDS—a disease unknown twenty years ago—is epidemic. Many of these ills represent problems of funding and distribution. Cholera is almost a trivial disease if patients receive antibiotics and clean water. Many do not. But nature also has proved more resourceful than doctors once imagined. Viruses previously unknown appear each year to strike us down. Despite scientific advances, we will still be battling these unpredictable killers in 2010.

In the developed lands, the average citizen enjoys conveniences not available to the wealthy of an earlier age. We eat nutritious food, drink clean water, benefit from modern medical care, drive air-conditioned cars, and tape our favorite television programs. Yet in the United States—the wealthiest of nations—malnourishment, infant mortality, and illiteracy are on the rise. These are matters of social policy, and technology alone is powerless against them.

What technology does promise is a better life for all whose economic and political situations allow them to take advantage of its wonders. And practical comforts it has delivered beyond any possible expectation. If we, on average, are better fed, live longer and healthier lives, enjoy a broader range of choice in both material goods and lifestyles than our ancestors could have conceived, science and technology deserve most of the credit. And so it will be in the decades to come. Technology is not the only power that will shape our future for good or ill. Yet it is the critical force that more than any other single factor, although not uninfluenced by the rest, will determine what is possible for us.

As forecasters, we spend most of our time writing and talking about the future. Unlike many in our profession, we hold to a relatively cheerful view of the years to come. Though humanity still confronts problems that have burdened it for decades or millennia, the early twenty-first century will be a time of relative peace and growing comfort throughout most of the world. Our future will be constrained by demographic, economic, and political forces that should be under our control yet seldom are—but it will be driven by our expanding knowledge of science and mastery of technology. In lectures, magazine articles, and several previous books, we have tried to account for all these influences.

Probable Tomorrows marks a change for us, for in this book we attempt simultaneously to narrow our focus and broaden our view. In the past, we have dealt with demography, economics, and politics. In this volume, they play only indirect roles; here we concentrate as best we can on the central theme of technology. In this realm, we encounter not only the confined future recognized by most political leaders but the vast expanses open to a humanity moved by vision and acting with courage.

Doing them justice has not been easy. Inevitably, any book about the future omits far more of its subject than it includes. This is especially true of a work that deals with technology, the most mobile facet of human endeavor and the one discipline whose entire purpose is to make the future different from the present. Merely to fit this survey into one volume, we have been forced to concentrate on relatively few topics, ignoring many others, most of which had valid claims for inclusion.

Computers obviously must be covered in as much depth as was practical. So too communications networks. Between them, these closely related technologies are transforming almost every field of human activity, from office work to medical care to the arts.

Medicine is another essential subject. Yet so many individual discoveries are being made almost daily that we have had to cull our potential topics rigorously. The few subjects that survived this winnowing nonetheless fill the largest chapter in the book, making it probably the most startling.

The environment also could not be ignored. Though information in this field often is equivocal or incomplete and forecasts are harder to make, new information is beginning to banish the uncertainties. And no subject could be more important.

Other cases were much less clear. For various reasons, we have slighted entire fields in which the near future is sure to bring sweeping change.

New technologies arrive first in war, where the Darwinian pressure for improvement is greatest and budgets are almost irrelevant. So it will be in the early twenty-first century. The line soldiers in the most modern armies of future conflicts may track their position in battle using satellite data projected onto helmet-mounted screens while state-of-the-art biosensors scan the air for biological and chemical hazards. Yet most future wars will be fought with obsolete weapons, by Third World governments and terrorist movements that seldom can afford the newest and best equipment, even if the great powers would trust them with it. The industrialized lands will take their high-tech arsenals to war only when they have such an overwhelming advantage over relatively backward adversaries that the outcome would be certain even if their soldiers were limited to M-1 rifles. Thus military technology becomes largely irrelevant to our lives.

Robots and automation will directly affect us all; yet we have largely ignored them. In manufacturing, computers already have made it possible to assign machines to most of the jobs once performed by human workers; over the next decade or so, they will send even more of us to the unemployment lines. At the same time, they will allow companies to tailor their goods for the individual customer, while retaining all the efficiency of mass production. Most products should therefore become cheaper (in real terms), even as their quality improves and consumer satisfaction rises. Yet these changes, which will touch all our lives directly, have as much to do with economic forces as with technology. New and more effective forms of automation will be adopted in the next decade or so, not because the world's R&D laboratories have produced revolutionary new manufacturing methods but because global competition requires them. In the end, we have decided to pass them by, save when they touch on one of our surviving topics.

Genetic engineering is another such case. It has already given us new and purer pharmaceuticals. Over the next fifteen years, it will provide still more drugs and diagnostic techniques; crops that grow in land contaminated by salt, flourish in arid climates, and resist insects and plant diseases; and bacteria that break down polluting chemicals into harmless materials. A good argument could be made for using some of this book's limited space to examine these developments. Yet as a technology genetic engineering itself has grown surprisingly mature in the last twenty years. In the end, its products seemed to fit best in other chapters, where they have been forced to compete with other developments for coverage. Few of them have managed to win out.

On the other hand, we have spent an entire chapter on transportation, where technology's promise is tightly constrained by government policy and investment priorities. To some extent, we chose to examine this field because the need to improve our transporation networks has become sufficiently urgent that political and economic obstacles could be overcome soon. And if, for example, some new kind of propulsion replaces the internal combustion engine, the effects will be obvious to us all.

Yet it is true also that in making these choices, we often simply have followed where our curiosity leads. Not all such decisions can be defended on purely rational grounds, or need to be.

This book tells of wonders that any latter-day Barnum might hawk alongside his acrobats and performing elephants. They are not the heady fantasies of science fiction but earthbound forecasts only a few short steps along paths whose direction is already clear. Most of these marvels will arrive as if on schedule, virtually uninfluenced by politics or government policy.

In just a few years:

- Personal computers will offer the power of today's supermachines and artificial intelligence yet unknown even in the laboratory.
- American high-tech companies and their competitors abroad will weave a telecommunications network that supplies the world with services from the contents of the Library of Congress to pornographic videos in Cantonese.
- The United States—reversing the trend of decades—will begin to link its major cities with high-speed railroads.
- Former defense contractors may bind the continents with single-stage-to-orbit aerospace planes capable of leaping halfway around the world in two hours.
- The industrial countries will use automation to mass-produce high-quality consumer goods at prices so low that the poor of tomorrow could live as well as the rich of today. The wealthy, of course, will enjoy even greater luxuries.
- Atmospheric scientists will have learned to purge the air of pollution, closing the Antarctic ozone hole and ending the threat of global warming (although they will not yet have had time to put their plans into effect).
- With imagination and commitment, humanity could begin to ship its heavy industries into space, so that Earth can recover from our past environmental follies. It will not happen in the next twenty years. Yet the development work now in progress should sharply

reduce the cost of launching material, or people, into orbit. And that could set the stage for more dramatic advances a decade or two later.

◆ Almost certainly, medical research will extend the healthy human life span beyond the century mark, and scientists will use that extra time to provide even greater extensions. Just conceivably, as a result of research now under way, some of us may never die.

Not all of these promises will be kept. Yet many will be. It is from among the broad opportunities brought to us by technology that we will, deliberately or by default, choose the one true future in which we all must live. If our forecasts err, it will be because our imagination and courage have failed us or because society has cast some opportunities aside and has chosen other paths instead. And the more we know about our options, the better the chance that we will choose wisely.

PROBABLE TOMORROWS

1

GET READY FOR DIGITAL EVERYTHING

"Future computers? Everybody in the business knows what's coming." The speaker was an engineer friend of ours who spends his days tinkering with next-generation prototypes. "In ten years, personal computers will be nothing like the boxes we use today," he said. "They will have the power of a supercomputer, but they will be no larger than a pack of cigarettes. You won't have to carry a keyboard or screen. Instead, just talk to the machine. Tell it what you want to know, and it will tell you the answer.

"Of course," he added, "by then, you could just wire the computer directly into your brain. But manufacturers have such a big investment in conventional technology that they'd go broke if they brought a neural interface to market that soon."

For special purposes, he offered two accessories, each of which would fit into a pocket. Those of us who are artists, designers, and writers all really want to look at our pictures or typescripts, not just talk to a machine about them. So a tiny projector mounted on a pair of glasses will beam a 3-D image directly into our eyes. Wherever we look, it will seem that a high-resolution color screen is hanging in space just where we can see it most clearly. And those who cannot quite free themselves from a keyboard will still be able to "type" and enter commands by pressing buttons. They will just have to carry an extra box, this one much like the remote control for a television. Even with these add-ons, the computer of 2007 will be one tiny package.

Our friend really does know what the computer industry is planning. He has spent the last twenty years of his life helping to develop salable products from the ideas handed down by its research departments. Yet we doubted about half of his forecasts.

For example, scientists have created computer chips that can be linked directly to nerve cells; yet that neural interface is a lot farther off than many optimistic engineers recognize. Before they can usefully wire a computer into our nervous systems, scientists will have to figure out exactly how the brain processes information. That has proved to be a lot more difficult than researchers once believed. And the surgeon who implants our new hardware will face liability problems that no insurance company would cover at any price. Someday, possibly, we will control our computers just by thinking at them—but not by 2007, or even 2017.

In contrast, that keyboard/controller is easy to build. However, we pity the poor marketing department ordered to sell it. Typing a single character on this little box requires pressing several keys at once—your index and middle fingers for "A," perhaps, your thumb and ring finger for "B," and so on. We have seen this concept before. In the mid-1980s, one company offered a single-hand "keyboard" that worked on the same principle. Engineers loved the idea. Customers hated the reality. Learn some weird new finger code? Most touch typists would rather graft a microchip to their brains.

Yet our friend's vision holds the essence of tomorrow's computers. Ten years from now, computers will be at home in our pockets—and inside our televisions and coffeemakers, and almost everywhere else as well. They will pack vastly more computing muscle into tinier, lighter packages. Compared with the dreams of some industry pundits, in fact, the forecast that began this chapter is remarkably conservative. If the most optimistic computer scientists are correct, tomorrow's shirt-pocket computer could hold a billion bytes in its working memory—about two thousand copies of this book—and run at 50 million times the speed of today's fastest personal computers. We have no idea what to do with all that computing power. We doubt that anyone else knows either. Even the lowest estimates of the advances coming in the next decade will give tomorrow's computers a level of utility and convenience we can only imagine today.

Computer engineers love acronyms almost as much as Pentagon bureaucrats do, and one of them has devised the perfect acronym for the future of computing: CADE. It stands for Convergence and Digital Everything.

Convergence is the grand coming-together of technologies that used to be separate. Radio and television and telephones used to work with "analog" signals. That is, they represented sound and images by varying the strength or frequency of their signals. (This is what the "M" of "AM" and "FM" means—"modulation," or varying.) Now they are going digital. Pictures and sound are broken into tiny pieces, and each fragment is represented by a series of computer bits. Just as CD systems give better sound than old-fashioned record players, digital TV gives clearer pictures than the conventional box that now dominates most living rooms. But a digital TV is not merely like a computer; it is a computer, with special software. And when you are not busy crunching spreadsheets or balancing your checkbook, your personal computer will be able to switch programs—in both senses—and display your favorite TV show. Your TV is a computer, and your computer is a TV. That is convergence. It is happening wherever technology meets information.

Already, the cheapest way to make a long-distance telephone call is to fire up your computer and log onto the Internet. Several companies provide telephone-style headsets for personal computers, together with the software needed to operate them. Using these devices, you can talk with any Internet user who has a similar headset anywhere in the world. At this early stage of development, the sound quality is variable—never up to that of a normal telephone but usually adequate. And the price can't be beat. One of the authors lives in southwestern New Hampshire, where he has an account with the more expensive (but more convenient) of two regional Internet service providers. At daytime rates, a conventional telephone call to Greenfield, New Hampshire—all of four miles from home, but not in the local calling area—costs $.26 for the first minute and $.24 a minute thereafter. Using the Internet, a conversation with someone in Greenfield costs $.005 per minute. So does a conversation with someone in London, Paris, Moscow, or New Delhi. The computer is your telephone, and your telephone is obsolete. Convergence.

This is just the beginning. For years, most of us have had three different sets of wires and cables entering our homes and offices. Electric wires carry power; telephone lines carry conversations and, increasingly, computer data; television cables carry . . . umm, entertainment?—video signals, at any rate. Recently, some electric utilities have added a fourth set of lines that allow them to check meter readings without leaving the office; most of us are not even aware these connections exist.

What happens when all the information is digital? Why not carry TV pictures on the telephone wires? Computer data on the TV cable? Or both of them on the electric utility's meter-checking lines? At least one power company already is renting unused capacity on its data lines to carry digital messages for local businesses. One recent proposal would even ship computer data across the power lines themselves. Each of these ideas offers its own benefits, and each requires the solution of some technical and economic problems. We examine them more closely in the next chapter. For now, just think of them as one more example of convergence.

Digital Everything is an even simpler notion. Computers are becoming so small, powerful, and cheap that soon almost any object more complex than pottery will be equipped with its own brain. Lights will adjust themselves to illuminate your book or keep glare off the CV (computer/television) screen. Toasters will learn whether you like your English muffins lightly browned or charred beyond recognition. Intruder alarms will know enough not to call the police just because you left your keys on the dresser, even if you have not upgraded to one of the new voice-recognition security systems. Edward Cornish, president of the World Future Society, recently completed a study of computer-driven trends in society. He cites the example of one Texas man who computerized his sprinkler system and programmed it to come on when errant golfers from the neighboring course trampled his lawn! The Digital Everything revolution has already begun.

Take a more complicated example: Imagine a stove that arrives with all the recipes from Irma Rombauer and Marion R. Becker's classic *The Joy of Cooking* and Graham Kerr's latest TV program stored conveniently in memory. A ROM-card reader will let you add recipes from new cookbooks as well. Just tell the stove what you want to prepare—like most computers, it will understand verbal instructions—and it will display a list of ingredients on its flat-panel screen. It will announce when the skillet is hot enough to sear a steak, prompt you when the pasta is al dente, give fair warning when the next step requires exact timing. It will sense when a soup is beginning to boil and automatically reduce the heat to a slow simmer. It will schedule all your meal's courses to be done perfectly at serving time. With use, it will remember how you like your food. If your vegetables come out a little too raw for your taste at the factory setting, the stove will learn to cook them a bit longer. Of course, it will learn different preferences for each cook in the family. No doubt it will have many other "intelligent" functions that have not occurred to us.

Fifteen years from now, product designers will still be figuring out startling ways to use the new intelligence of everyday appliances. No one of these innovations will change our lives. But as the artifacts around us gradually learn to accommodate our individual needs, the world will become a friendlier, more convenient place in which to live.

We return to some of the more promising uses of tomorrow's computers later in this chapter. For now, let us look at the technologies that will make them possible.

When it comes to microprocessors, even our minimum expectations are astonishing. In fifteen years, these chips will be about 15,000 times more potent than the processors that power today's cutting-edge personal computers. Tomorrow's run-of-the-desktop computer will finish in an hour a task that today would keep our most powerful desktop computers running twenty-four hours a day for two years. This seems even more amazing because it requires no revolutionary technological breakthrough. The futuristic superchip of 2010 will be much the same mass of transistors etched into silicon that engineers have been improving since the first commercial microprocessor, the Intel 4004, appeared on the market in 1971.

The difference lies in greater component density and more sophisticated design—the same two factors that make today's advanced microchips so much more powerful than the primitive processors of the 1970s. That first Intel 4004 packed some 2,300 transistors into a space roughly the size of a fingernail. The Intel Pentium Pro, currently the firm's most powerful chip, contains 5.5 million transistors in a space not all that much larger. There are a lot more circuit elements to do the work, and they are packed more closely together, so the data can move between them faster. This is one reason the 4004 could carry out one program instruction 60,000 times per second, while the P6 can perform 250 million instructions per second. (The fastest experimental processors made today are four times faster yet.)

The other reason is a matter of tactics. A computer program consists of an enormous list of minute steps: Bring one number from external memory into the processor; bring in another number; move one of them from its temporary storage register into the number-crunching area; add the other number to it; move the answer into another storage register; look up the programmed instruction for the next step; and so tediously on. Early computer chips processed these instructions one at a time until the program was complete. It was a lot like trying to move the entire population of New York City through a single subway turnstile.

In recent years, computer engineers have worked out ways around that bottleneck. Modern microprocessors are "pipelined" and "superscalar." Pipelined processors move several instructions through the system at once, opening more turnstiles. A single stage of pipelining halves the time it takes to work through a program. Current processors are nearing ten stages of pipelining, and their descendants will have many more. Superscalar processors can perform several instructions at once, in effect stuffing several people through each turnstile at the same time. Again, this multiplies the machine's effective processing speed. Current processors carry out three to six instructions at one time. In fifteen years, the number is likely to be several dozen. These incremental advances alone will make tomorrow's computers several hundred times more powerful.

We are less optimistic about another strategy from which researchers have long expected much greater advances in computing speed. This is parallel processing. This technique aims to break a problem into many smaller tasks, perform each one simultaneously on its own processor, and then recombine the results of the individual computations into a single answer. In theory, this should be the ultimate upgrade.

At this point, we have worn out our subway analogy. Instead, imagine adding two very large numbers. We add the ones column, carry a number, add the tens column, and so on. A parallel processor might tackle this simple problem by adding all the columns at the same time, each on its own subcomputer; still more processors keep track of the carries. In the end, the machine combines the results of all these separate calculations into the final sum. Nothing could be faster, assuming that it can be done at all.

It turns out that making such a computer is not terribly difficult. Engineers have built parallel-processing computers that contain hundreds of identical microprocessors; a few small companies even specialize in making this kind of machine. Computers with many thousands of processors already are on the drawing board. For certain specialized kinds of calculation, nothing else approaches the speed of parallel processors.

However, distributing each program among many processors has proved to be almost as hard as passing one of our New Yorkers through several turnstiles at once. It is difficult enough to separate most computing problems into easy-to-process fragments, and that is only the first hurdle that programmers face. Distributing those many parts to individual processors, keeping the subcomputers in step with each other, and then reassembling their answers into one grand result has proved all but impossible, save for those few specialized chores that lend themselves to subdivision. These problems will not be solved until someone achieves the kind of conceptual breakthrough whose appearance no one can predict. We suspect that programmers will still be struggling with them fifteen years from now.

While they do so, engineers will overcome some obstacles that make it difficult to continue cramming ever more transistors into silicon little larger than a postage stamp. It will not be easy, because the circuit elements of a microprocessor are so tiny already. The smallest structures on today's memory chips are only 0.35 micron across, or 35 hundred-millionths of a meter, just over one-three-hundredth the diameter of a human hair. By 2000 or so, they will be 0.1 micron across, or roughly the width of a coil of DNA.

Microchips are made by photolithography. Though the process has become extraordinarily complex, in practice it is simple, much like printing an ordinary photograph. The printing machine, known as a stepper, shines light through a mask onto a wafer of silicon, which is coated with a material called a photoresist. The mask is a simple black-and-white picture of the circuit being created. (Today's masks are made of chromium deposited on quartz.) Wherever light penetrates the negative, the photoresist hardens; where the negative blocks the light, the resist is unchanged. Unhardened resist is then washed away, leaving bare silicon. These exposed areas are etched with acid to create the basic circuit, coated with metal to form conductive areas, and treated with a variety of other materials that alter the silicon's electrical characteristics. A single chip can contain as many as twenty layers of circuitry, each one laid down by this exacting process.

Those circuits have become so small that photolithography is hard-pressed to make them. The waves of light itself are as large as the components, so they cannot be focused sharply enough to produce a usable image on the silicon. The traceries of photoresist that adhere to the nascent chip are too bulky as well, and when made smaller they tend to flake off the silicon. Over the years, chip makers have moved from using visible light to ever-shorter wavelengths of ultraviolet, and they have developed new resists able to capture ever finer circuit details. These refinements have been a major factor in the development of today's densely packed microchips, but they have nearly reached the end of the line. In the next generation of processor chips, or at the latest the generation after that, the components will be so small that radical changes will be required to make them. For state-of-the-art chips, photolithography soon will be obsolete.

Several alternatives will come to the rescue of chip designers. X rays are far smaller than light waves, so they can make much tinier components. Most chip manufacturers now believe that X-ray lithography eventually will move from the laboratory to the factory clean room. Late in 1994, a consortium of American chip manufacturers joined forces to develop this technology. Another possibility is to etch each line of a circuit directly into the silicon with a beam of electrons, protons, or helium ions. It is a promising idea, but thus far it takes too long to steer the beam through the intricate patterns of a microcircuit; making a single chip this way can take hours. Harvard University physicist Mara Prentiss and her colleagues AT&T's Bell Laboratories (now Lucent Technologies) have even used light to steer individual atoms onto the surface of a silicon chip. This has raised hopes that they eventually will be able to scribe circuits one-tenth the size of the smallest now available. It is too early to say which of these technologies will be used to manufacture tomorrow's advanced microchips. However, it seems certain that one of them will be available for use when it is needed.

There is one more way to make faster microcircuits: Instead of compressing them still further, change their properties so that electrons move through them faster. This means replacing "old-fashioned" silicon with some other material. For years, the favorite candidate has been gallium arsenide (GaAs), a semiconductor that makes possible computing speeds several times as fast as silicon can offer. However, switching to GaAs requires an enormous investment in specialized equipment—overwhelming even by the standards of an industry in which the price of a new factory now starts at over $1 billion. To date, chip makers have been unwilling to make that kind of commitment. Instead, companies such as IBM and Analog Devices have been working with an alloy of silicon and germanium. Alloy circuits can operate twice as fast as traditional silicon chips, and they can be made on the standard equipment that manufacturers are already using. A few high-performance alloy chips are already in production. At this point, it seems likely that state-of-the-art microprocessors of 2005 will include a little germanium; but they will still be composed largely of the same old silicon from which their earliest ancestors were made.

BREAKTHROUGH TECHNOLOGIES

In mid-1995, the U.S. Department of Energy commissioned Intel Corporation to build the fastest computer yet: a parallel-processing machine that will link more than 9,000 of the company's top-of-the-line

Pentium Pro chips. The new machine will be the world's first "teraflop" computer, capable of 1 trillion floating-point operations per second. (A floating-point operation is a computational step that involves decimal fractions, which are considerably slower to process than integer arithmetic.) This is an astonishing speed, at least for those of us accustomed to garden-variety desktop computers. This next-generation bit cruncher will be six times faster than today's quickest experimental computers.

However, even that will not be fast enough for some purposes. Such rarefied applications as image processing and recognition; modeling chemical processes, global weather patterns, or the stock market; orbital mechanics; particle physics; and cryptography require even greater speeds. This is the realm of the "petaflop" machine, capable of 1,000 trillion floating-point operations per second. An informal research group led by NASA scientists is already brainstorming a conceptual design for the petaflop machine. Current schemes call for a parallel-processing computer made up of 10,000 superconducting microprocessors, each of them carrying out 100 billion floating-point operations per second. It could be ready for use—with a price tag of between $100 million and $200 million—by 2015.

On the other hand, even these super-and-then-some computers may already be obsolete. One entirely new technology has already shown that it can reach computing speeds far beyond anything mere petaflop machines have to offer. A second form of future computer might not be quite so fast—it is not clear yet just what it is capable of—but it might well handle certain kinds of problem more efficiently than any of its competitors. And while a third innovation may yield computers that are only ten times faster than the fastest of today's laboratory marvels, there is a good chance it could be built at a price that would put one in almost any computer user's pocket.

The newest and most revolutionary breakthrough in computing burst onto the scene entirely without warning. One night in the summer of 1993, a computer scientist named Leonard Adelman picked up a little light bedtime reading, James Watson's classic textbook, *Molecular Biology of the Gene* (Addison-Wesley, 1987). As he read himself to sleep, Dr. Adelman was fascinated by the parallels he saw between genes and the computers he dealt with each day. Computers manipulate information that is encoded as binary numbers, strings of zeros and ones. The genes, he saw, manipulate information that is encoded as strings of adenine, cytosine, guanine, and thymine, the four nucleotides that combine to form DNA. Then he had one of those moment of illumination that sometimes come to scientists who are smart, prepared, and lucky: The DNA computer was much more than an intriguing metaphor. He could build it.

By Christmas, Adelman had worked out the details. He had also decided to test his new computer on a difficult mathematical challenge known as the traveling salesman problem: What is the shortest route among a group of cities, not all of which are connected by a direct road, that lets a salesman pass through each city exactly once? Given only a few cities, the problem can be solved with a pencil and paper. But as cities are added, the number of possible routes explodes. If a teraflop computer had begun to work out a 100-city traveling salesman problem on the day the universe was born, it still would not have found the answer. Adelman chose to work with seven cities interconnected by fourteen roads. To make the problem a bit easier, he chose the cities in which the journey would begin and end. His salesman would travel from Detroit to Atlanta, with stops in five other cities. It was still a daunting task.

Adelman assigned a code to each city and to each possible leg of the trip, twenty-one unique sequences of nucleotides to represent all the factors involved in his problem. Then he mixed the snippets of DNA in solution and allowed them to react for a few days. To read the answer, he simply analyzed the resulting DNA. The shortest strand that began and ended with the appropriate cities and contained the codes for all the intermediate stops represented the answer. On its first try, the DNA computer solved its problem faster than any electronic computer could have done. As a bonus, it even used less energy.

None of this means that DNA computers will soon flood into your local Bytes 'R' Us. For a simple problem, the nearest PC probably is faster than Adelman's wonder. The DNA computer is at its best when dealing with huge calculations that can be handled conveniently by trillions of processors—the DNA molecules—all working in parallel. And DNA is prone to errors; it carries our genetic code faithfully only because the cell contains extensive repair mechanisms to weed out defects. Worse yet, it took the DNA computer only moments to come up with its answer—but it took Adelman a week to purify the DNA that held the salesman's route. This is significantly more tedious than reading the answer from a color monitor.

There was one other obstacle as well. Adelman's prototype computer worked only because he had designed his DNA codes very cleverly to represent his version of the traveling salesman problem. It could not have carried out any other computation or even another variant of the traveling salesman problem. Some critics held that the DNA computer could never become a general-purpose computing machine. However, less than a year after Adelman demonstrated his prototype, a Princeton computer scientist named Richard Lipton figured out how to represent the zeros and ones of an ordinary computer as patterns of DNA. He also worked out a means by which a DNA computer can carry out all the logic operations possible for its electronic competition. In principle, there is no longer any reason Adelman's invention cannot become a general-purpose thinking machine.

At this point, it seems that DNA computers could replace all those massively parallel supercomputers before the petaflop machines are even built. Then again, there is still that problem of reading the answer without spending a week at the nucleic acid analyzer. So far, this technology is too new for us to make any firm prediction about where it will lead. In five years, we should have a good idea of how well it is working out.

Scientists have been toying with the idea of quantum computers, the second class of miracle machine, since the early 1980s. Yet only in the last few years have they begun to take them seriously. Computer designers first started thinking about quantum theory when they realized how their circuit elements were shrinking. Make the transistors on a micro-chip small enough, and strange things start to happen: For one, electrons disappear from your memory circuits and reappear somewhere else, without ever having existed anywhere between the two positions. It was a disturbing thought for engineers whose machines required exact con-trol over the electrons that powered them. But theoreticians wondered whether there might be some way to harness these strange phenomena.

In theory, there is. We will not get into the details here. Niels Bohr, the Danish physicist who was a leading pioneer of quantum theory, once commented that "Anyone who can contemplate quantum mechanics without getting dizzy hasn't properly understood it." Nonetheless, phys-icists with a high tolerance for vertigo have worked out ways in which quantum devices could carry out all the basic logic operations required of computers. In the mid-1990s, a few of these functions have even been realized in the laboratory. Again, there are many technical hurdles to be overcome before these proof-of-concept experiments give rise to work-ing quantum computers. And yet . . . and yet it now seems possible that someone, someday will produce a useful machine based on quantum theory. A decade ago, not even the theoreticians held out much hope.

What good would a computer be if you can't even be sure whether its bits are ones or zeros? It turns out that one of the easier possibilities is to simulate other quantum systems. According to one of this field's leading theoreticians, Dr. Seth Lloyd, of the Massachusetts Institute of Technology, a quantum computer capable of tracking only 40 bits of data through 100 logic operations could perform simulations that would take years on the largest supercomputers now available. It could revo-lutionize some fields of physics. Here in the practical world, a much more powerful quantum computer could easily break the government's "Clipper" encryption system and related public-key ciphers, including those used to protect electronic bank accounts. In the long run, quantum computers might turn out to be far smaller and more powerful than anything based on more conventional technologies.

Quantum computers remain a long way off, however. In the next ten years, physicists might just create the 40-bit computer of their dreams. We doubt that they will produce a reliable general-purpose quantum computer for at least a decade thereafter. It could take much longer than twenty years.

In sharp contrast, optical computers—the third breakthrough technology—could well be in general use just ten years from now. To date, most work in optoelectronics has gone into CD-ROMs, flat-panel displays, high-speed telecommunications lines and switching systems, and sensors for digital control systems—items related to computers, but not the core technology. Now computers themselves are going optical.

The benefits are overwhelming. Optical computers will be faster, because light moves so much faster than electrons mired in silicon. This alone is such a powerful advantage that in the early 1990s, four computer companies teamed up with the Pentagon's Advanced Research Projects Agency (ARPA) to create an optical backplane, the wiring that links a processor with its memory chips and communications ports. Just hurrying the transfer of information to and from the microchip, they felt, would provide a dramatic increase in computing speed, even without an improvement in the processor itself.

Optical computers also offer a radical change in the way information is handled. Electronic computers work with single bits of information. Optical computers can process whole images—potentially billions of bits—at one time. For image processing especially, the result is a radical increase in computing speed. Several years ago, computer scientist Kristina Johnson of the University of Colorado designed an optical processor that could identify cancer cells on a Pap smear. For a standard electronic computer, this task would have required a major commitment of processor time and memory, and it still would have meant a lengthy wait for the results. Johnson's dedicated optical processor did the job virtually instantaneously.

Like many other advances in technology, converting entire computers to work with light has proved more difficult than scientists once imagined. All-optical computers still are not much more powerful or practical than the prototype, which was built at AT&T's Bell Laboratories back in 1990.

However, scientists at Lehigh University recently have made an important advance. Alastair McAulay and his colleagues have built a high-speed optical memory unit capable of permanently storing vast quantities of data. In theory, the device can perform all the functions that now require a hard drive, costly RAM chips, and even the microprocessor's own onboard cache memory. This development alone promises to transform the computers we know today.

If Dr. McAulay is correct, computers soon will rely on optical circuits for vast memory and high-speed internal communication and retain electronics for what it does best—hard-core number-crunching in the microprocessor. In just ten years, these mostly optical computers could operate at clock speeds of 10 gigahertz, ten times faster than today's supercomputers and fifty times faster than the quickest PC. A typical machine would have 1 billion bytes of working memory—enough to hold 50,000 double-spaced pages of typing—and another 100 billion bytes of mass storage. Its flat-panel color screen would offer resolution about twenty-five times better than a good 35-mm slide. As a bonus, it could operate for weeks on a set of batteries.

We doubt that Dr. McAulay's vision will materialize in the next ten years, if only because it would force computer companies to discard an enormous investment in the factories that make memory chips and hard drives. But optical backplanes and other specialized components almost surely will have arrived, and mostly optical computers will be nearing the market. The result could be the shirt-pocket computer our engineer friend expects, but with even greater power.

MAKING IT WORK

We could easily fill this book with hardware technologies that promise to improve computers over the next ten years. New memory systems will pack vastly greater quantities of data into ever smaller volumes; the only question is which of many competing technologies will win out in the marketplace. More sophisticated signal processing will tease the meaning from noise-filled sounds and communications. Flat-panel screens finally will provide high-quality color images in sizes now available only from bulky TV-tube monitors. And so on, almost without end. Yet the time has come to move on.

Without software to control it, all this hardware is just scrap metal, plastic, and highly purified sand. That means there is a lot of programming to be done in the next ten years. In this area, we see only two major innovations that are likely to pay off in the early twenty-first century. One is a child of the 1960s. The other has an even longer history.

Object-oriented programming (OOP) systems are the Lego blocks of the software world. Though high-tech in detail, in spirit they are as simple as a brick wall. In traditional programming, writing software is a tedious, tricky, repetitive job. Each segment of every new piece of software is written from scratch. Programmers may build up their own personal experience, but they never create the kind of heritage of reusable parts that mechanical engineers and architects rely on. If a new project must send information to a printer, someone must write a printer driver for it, no matter how often programmers have written similar drivers in the past. On average, a good programmer turns out perhaps ten lines of finished code in a working day—this in a project that may run to millions of lines. In contrast, OOP dictates that programmers break their creations into reusable modules, or "objects." Then, when the time comes to write yet another piece of software, all the programmer has to do is pick the right modules off the shelf and knit them together. Instead of creating the entire project from a blank page, the programmer need only write the links between these objects and perhaps a few lines of highly specialized instructions unique to that project.

The potential savings in time are obvious. Steve Jobs, the marketing half of the two-man team that designed and built the first Apple computers, is so taken with the idea that he has transformed his NeXT Computer, Inc., from a struggling hardware manufacturer to a software shop specializing in object-oriented programming. He estimates that OOP will cut 80 to 90 percent off the time it takes to develop new computer applications.

There is another benefit as well. Every time someone writes a line of computer code, it is an opportunity to make a mistake. And programming errors can be devilishly hard to find and correct. In developing any major new product, software companies spend months or years ridding their creation of programming errors—and almost inevitably the product reaches customers months late and still plagued with undiscovered problems. More than one company has lost market share because the long-awaited new edition of a key product was so buggy that customers made the difficult move to competing software. With OOP, debugging should be relatively quick and easy. Once each program module has been perfected, it can be used in future projects without problems. This means that even in the most complex new program, a large majority of the code is known to be correct. Only the project's all-new code must be debugged. And this in turn means that the product will reach the market months earlier, customers will be able to use it without problems, and future programmers assigned to upgrade the software will have a much easier time of it.

There is nothing really new about OOP's modular approach to software creation. Back in the 1960s, when programmers at AT&T wrote the code that still runs America's telephone system, the software was broken into reusable blocks. The idea then was not so much to reuse the code as to make it more reliable and easier to maintain. The concept worked. This is one reason the United States enjoys some of the best telephone service in the world. IBM used much the same approach in creating the operating system for its classic 360 mainframe computer. At the time, it was one of the largest programming projects ever devised. Programmers since then have made halfhearted attempts to subdivide their software into reusable blocks, but their efforts are only just beginning to pay off.

In the next fifteen years, the demand for new software will grow with each new hardware innovation. The supply of experienced programmers will grow too, but much less rapidly. Programming might easily become the kind of bottleneck that would cripple the development of useful products from all the bright new ideas emerging from the world's research centers. OOP will keep the computer revolution moving at full speed well into the next century.

The second key technology is artificial intelligence, or AI. It was one of the hottest computing topics of the 1980s. Hal, the deranged computer who served as the villian in the movie *2001*, was sentient, almost alive—and he was already more than a decade old. In a plausible future, which it seemed we were fast approaching, computers would hold conversations, go about their jobs with only a minimum of oversight, understand the sometimes vague notions that guide humans through an imprecise world. In the 1980s, it seemed the machine intelligence of *2001* might arrive even before the year did.

Yet AI has fallen out of favor in the 1990s, largely because computer scientists promised more than they could deliver. Researchers founded literally dozens of trendy little companies to develop marketable products incorporating this sexy new technology. All they delivered was some limited programs that "canned" human expertise in a series of IF-THEN statements: "IF the Fed cuts the interbank lending rate, THEN the Dow-Jones Industrial Average will rise for the next week. (Confidence 80 percent.)" "IF the applicant's total credit card debt exceeds annual income THEN the loan application should be rejected. (Confidence 95 percent.)" Several thousand of these so-called expert systems are now streamlining business and engineering procedures from detecting credit card fraud to picking winners at the track. These are useful abilities, but not the stuff of technological revolutions. Most of those hot new AI companies born during the early 1980s died before their decade.

In the years to come, AI will get a second chance. One reason is the growing power of computers. It takes a lot of memory and processing capacity to mimic the human mind, particularly when it often must be done by overwhelming complex problems with sheer computational power rather than by actually understanding how the mind works. Now that power is fast reaching the end user's desktop. Ten years from now, we all will have computers potent enough to incorporate very sophisticated forms of AI.

Another factor is the work put in by all those researchers in the 1980s and their descendants today. We still do not understand the mind well enough to duplicate many of its functions. Thus far, no computer can recognize one face among a crowd, a task we perform almost instantly; we do not really know how we do it ourselves. Over the years, however, scientists have built up a fund of knowledge that, given sufficiently powerful hardware, could finally lead to practical AI.

The third motive is simple necessity. To date, in order to make their products more versatile, software companies have made programs more complicated. Look at the manual for any state-of-the-art word processing program, and you will see the problem. Word processing is one of the simpler tasks you can ask a computer to handle. Yet these programs have grown so complex than it can take weeks to become even marginally comfortable with a new package, and few users ever really master the word processor they use every working day. At this point, software companies are beginning to recognize that their continued growth depends on selling to potential customers who will not put up with that kind of inconvenience. Future products must be both powerful and easy to use. They must take orders from people who are not interested in learning complex commands or even in using a mouse to pick them from a menu. They must be, in short, intelligent.

There are two ways to create machine intelligence of a high order. One is to give a computer a context, the kind of background information that enables people to interpret new data. The other is to give it the ability to learn. Researchers are hard at work on both approaches.

At the moment, the master of context-based AI is Dr. Douglas B. Lenat, who began his research at the prestigious Microelectronics & Computer Technology Corp. and recently has founded his own company, Cycorp, to continue his development work. Lenat's contribution to AI is CYC (pronounced "psych"), an experimental program that seeks to endow machines with the most elusive of human traits, common sense.

Common sense, it turns out, is largely a matter of background information, the shared fund of wisdom that most people learn simply by living and assume that others also will know. For example, when we learn that one woman is another woman's mother, we automatically understand that she is the older of the two. We recognize this without having to think about it, because we already know a lot about mothers and daughters. In particular, we know that one of them must have existed before the other was born. This is a subtlety that most computers would miss. Even if machines are programmed to draw inferences, they do not know enough to arrive at the right ones.

For a computer, CYC has an amazing fund of background information. Where people would have learned things through experience, Lenat and his colleagues have been working since the early 1980s to identify useful information and incorporate it into their program. CYC knows that people who fall may be injured, that people sweat when they exercise, that concerts usually involve musical instruments. To date, CYC has learned roughly 1 million such rules; Lenat aims eventually to give his creation perhaps 100 million such items of conventional wisdom to help it interpret the world as human beings do.

In a typical test, Lenat told CYC to find pictures of adventurous people. The program sorted through stored captions and found a picture of a man climbing a rock face. CYC's reasoning: Rock climbing is a sport. Climbers might fall. People who fall may be injured. Someone who risks injury in a sport is adventurous. Not all tests are so successful, but CYC is improving at this very human kind of interpretation.

At the Massachusetts Institute of Technology, Rodney Brooks has taken the other approach. Brooks is best known for his work with small robots that use networks of simple programs, each running on its own microprocessor, to mimic activities that appear far more complex. For example, one of Brooks's robots is a small, insectile contraption with six legs. Each leg has its own microprocessor to control its movements. One theoretical layer up, another bit of intelligence coordinates the motion of the legs. Specialized programs take care of functions such as climbing over obstacles or plotting a path around them. At first, the machine waved its legs spastically. But within minutes, it learned to walk.

Brooks believes that the same approach can help machines learn other human functions. His test case is a robot named Cog. With a head, torso, and a single arm but no legs, Cog is a few levels short of the classic science fiction humanoid. Yet in the long run, it could become very human indeed. Separate programs, each running on its own processor, will govern the movement of its head, neck, eye—a video camera, really—and arm. Other subsystems will recognize faces, censor harmful or self-destructive behavior, absorb new information, and provide overall control. If all goes well, each of these operations will get better with practice. Brooks is convinced that learning in this fashion eventually will let his creation mimic human behavior in ways far more complex than any rule-based system can master.

Being only human, Lenat and Brooks are slightly bitter rivals for dominance in the field of AI. But our guess is that their work will converge in the end. In the future, and probably sooner than either scientist would expect, computers will combine vast stores of context with dedicated learning software. Our machines will behave in ways that seem human, if sometimes a bit stilted and too literal-minded. If we give them an ambiguous or open-ended order, they will interpret it according to their knowledge of us. More often than not, they will get it right, and as they gain experience, they will do better still. In fifteen years, we may already have begun to take these intelligent, helpful, and artificial companions for granted.

WHAT GOOD IS IT?

The real question is, what are we going to do with all these new electronic abilities? Scientists may need a supercomputer's power to simulate the interactions of atomic particles or the results of a novel chemical reaction. The telephone company, or whatever it evolves into when the Internet dominates long-distance communication, may need twenty-first-century supercomputers to keep its lines operating. Moviemakers will use all the power they can get for lifelike animation and to resurrect the stars of another age. ("What about casting Monroe opposite Travolta? Plot? Who needs a plot with talent like that? Let's do lunch!") But what do the rest of us get out of all this new technology?

Entertainment is one obvious answer. The media now promise us, or perhaps threaten us with, 500 channels of cable television, or its digitized equivalent. This should be enough to keep even the most ardent channel surfer perpetually busy in front of the tube. Already, the Internet's hobbyist newsgroups and Web pages absorb a large and growing part of computer users' leisure time; in 1996, advertisers unhappily noted that all those hours in front of the computer were beginning to eat into their customers' TV-viewing time. And thanks to the growing power of personal computers, we soon may begin to make much more of our own entertainment. Computerized sound synthesizers already are popular among technologically savvy adolescents with a taste for music. Soon any desktop computer will be equipped for lifelike animation and video editing. Just install the necessary software, and the average teen will be able to create digital "films" that would challenge today's professionals. We know several who would love to have that power now.

The new computers will bring chaos as well, just as their predecessors have done. In the past twenty years, assembly-line robots have displaced well over half of the human workers who once looked to factories for a middle-class living. They have destroyed the jobs of countless mid-level executives, as computerized management methods allow the survivors to oversee up to twenty-one subordinates instead of only six. And they have made it increasingly easy for one company to absorb its rivals, sending still more people to the unemployment lines. (Despite occasional rumors of a possible merger between Apple Computer and Sun Micro-systems, we will not be surprised if IBM eventually absorbs its most innovative competitor.) Recently, computers have begun to streamline operations in the service industries, the last stronghold of human labor. Again, the opportunities for human workers are shrinking.

Cheaper and vastly more powerful computers can only hasten this trend. Who needs to pay a human salesperson when business customers can e-mail their orders directly to the computer that does your account-ing and controls your production equipment? Or when intelligent ma-chines can make an effective sales call to new retail customers? Who needs human doctors when the next generation of expert systems will be able to field most diagnostic problems? Even writers and artists may be in danger. Already some Hollywood screenwriters rely on their soft-ware to construct salable plot outlines, and one experimental program reportedly can write believable dialogue by reweaving fragments ab-stracted from real conversations. For a time, the health care industry can absorb many of the people displaced by this new, more capable brand of automation. After all, sick people still need human hands to tend them. But as we will see in Chapter 9, soon fewer people will be sick enough to require their care. Most of us will spend our entire working lives bouncing from one career to the next, scrambling to learn the skills of a new profession before some computer snatches our current living from beneath us. Chaos indeed.

Yet, almost paradoxically, what we really expect from the enormous new power of tomorrow's computers is convenience. It will appear in computers themselves first. Cycorp's Douglas Lenat forecasts that much of the software we struggle with today soon will develop common sense. Word processors will understand our text well enough to notice that we promised to discuss four major topics but mentioned only three. Spreadsheets will recognize that a calculation can be performed but does not make sense in the context of what we are trying to accomplish. So-called intelligent agents will hunt for information on the Internet, schedule our appointments by negotiating with our colleagues' agents, and perhaps even handle our routine shopping.

One of those agents will inhabit your "television." The saving grace of all those new cable channels is that we will never have to watch them or even scan a *TV Guide* the size of the Manhattan telephone directory. Our computerized, intelligent television will watch all those channels for us. It will also keep a continuous watch on the Internet, scanning the newsgroups and Web pages we prefer and occasionally adding a new service to its list. And it will know what interests us. Whenever we check in with the machine, it will deliver our e-mail, flag interesting items from the Net, and offer a short list of television programs it has stored for possible viewing. If some new topic has caught our attention, we will be able to ask it whether it knows of any programs or Net sites dealing with that subject. Like all our other appliances, this media butler will learn and grow continuously better at meeting our personal needs.

Telephones will survive, not so much because we need them—their functions could easily be built into the media butler—but because we find it easiest to compartmentalize some operations. The growing power of computers will change them as well. It will not matter much whether telephones are portable or hard-wired. They will route calls over the same network of wires, fiber-optic cables, and satellites that carries our data—whatever gets our message to its destination most efficiently. Wherever we go, whichever telephone we use, our personal phone number will follow us, much as calling card and other special-service numbers do today. The telephone itself will filter out unwelcome callers, intercept calls when we are too busy to receive them, and deliver special messages when it recognizes a friend's voice on the line. The phone system will perform one more service as well: It will translate our conversations in real time, so that we can talk with someone who speaks only French, Japanese, Mandarin Chinese—any of the seven or eight most common languages—without bothering to learn anything but our native tongue. Most of these chores are theoretically possible even with today's technology. In ten years, the growing power of computers will make them practical.

All of these functions, and many others, will be united by a local area network built into your house or apartment and—if you do not yet work at home—able to link itself with the network at your office. Your media butler, lighting system, and appliances will not just contain their own intelligence, cut off from the rest of the world; they will share information. When you sit down to read in the den, the lights will automatically focus on the page and slightly dim the rest of the room. They will also notify your telephone where to find you, so that when you get a call only the nearest extension rings and no one else in the house is disturbed. Or, if you prefer, it will tell the phone to hold all calls. Check in with the media butler later, and it will mention that the telephone has a message for you—or deliver the message itself. And so on. Nicholas Negroponte, founder of the famed Media Lab at the Massachusetts Institute of Technology, once suggested that if you want yesterday's closing price on the Dow-Jones Industrial Average served up with breakfast, you could have it automatically branded on your toast. No doubt he was half joking, but an intelligent toaster linked to your media butler could do it easily. As they learn what we find most convenient, these increasingly brainy artifacts will automatically find ways to insulate us from the endless inconveniences and irritations that have become inescapable parts of life in the late twentieth century.

The late Allen Newell, an influential pioneer of artificial intelligence, likened the future to the land of Faery, a magic place with friendly, if nonhuman, companions on all sides. We see the early twenty-first century as a kind of Walt Disney Wonderland, in which our possessions talk to us and to each other. But unlike the Mad Hatter's teapot, they usually will make perfect sense. And unlike Hal, of *2001: A Space Odyssey*, when they act behind our backs they will be plotting to do us good.

2

REACH OUT AND TOUCH . . . ANYONE IN THE WORLD

Early in the morning of January 17, 1994, the ground beneath the Los Angeles Basin trembled for less than a minute. Bridges collapsed. Buildings fell. And thousands of computer-literate Americans showed the rest of us what it is like to live with access to the Net, the global computer network that is now penetrating into homes throughout the developed world.

For twenty-four hours after the earthquake, Pacific Telesis blocked incoming service to make it easier for residents to call out and contact those outside the afflicted area. But for users of the Internet, Compu-Serve, and other computerized database and messaging systems, that day was almost business as usual. They simply turned on their computers, dialed up their favorite online bulletin board, and e-mailed requests for information to users in Los Angeles. Some 20,000 people on the Prodigy system alone sought and received word of roughly 100,000 friends and relatives long before normal phone service was restored.

By the turn of the century, or very shortly thereafter, most of us will have access to computer services even more powerful than the best now available. Almost anyone who now owns a telephone or cable television can or soon will be able to tap into online libraries, news services, and shopping malls; play multiuser virtual reality games; obtain up-to-the-minute stock prices; and make use of almost any other computerized service available anywhere in the world. Today's network services—CompuServe, America Online, Prodigy, and their competitors—are dim foreshadowings of the Net.wonders to come. In the long run, virtual libraries may well contain all the knowledge amassed by a curious humanity over the millennia of civilization. Much sooner, those of us linked to the Net will be able to:

- Watch almost any film ever made whenever we have the urge to see it;
- Tune in any of perhaps 500 channels of cable-style television;
- Have a self-directed computer program, called an "intelligent agent" or "knowbot," shop for the best price on airline tickets and book us aboard a convenient flight to our destination;
- Have another intelligent program cull stories on topics of interest from global news services and assemble them automatically into a personalized "newspaper," to be read on screen or printed out at will;
- Wander through an online "mall" and buy anything from clothes to septic tanks, viewing them with photographic clarity;
- Call up traffic data for the trip to work—assuming we do not yet work at home, as many of us will—and view the directions on an electronic map while sitting in the car;
- Join distant colleagues in an impromptu conference, passing around references and sharing data as though in the same room
- Search through the card catalogs of the Library of Congress and other major institutions throughout the world;
- And read the full texts of most books, magazines, and professional and academic journals published anywhere in the world.

This may not sound like a revolution in the making—not everyone finds it easy to be excited by the prospect of 500 channels' worth of pap and pablum—but keep reading. The techno-pundits are right: In the not-so-long run, new information services will change the professional and personal lives of virtually everyone in the developed world.

In the next few pages, we will try to forecast both the development of the Net and the changes it will bring. It will not be easy, because many of the critical decisions that will shape the result are still being debated. Relatively few involve technology; many are legal, political, and social.

A QUICK INTRODUCTION

Short of hiding in a cave for the last few years, it has been almost impossible to avoid hearing about the global data network. Nonetheless, a little background material may make it clearer just what it is that is beginning to tranform our lives.

The Net, as it is informally known, came into being as a classic Cold War artifact. During the 1950s, when relations between East and West were at their chilliest, it occurred to Pentagon planners that a thermo-nuclear war could destroy military communications systems, making it impossible to fight effectively, even if the necessary weapons and personnel remained intact. There was just no effective way to protect the giant command centers, antennas, and wiring that carried messages through the military hierarchy.

The solution to this problem was found in a scheme called packet-switching. Each message would be broken into small packets of data, each labeled with the addresses of its origin and destination and with information about which packets it should be connected with to reconstruct the entire communiqué. The packets would be sent out almost at random into the network, and routing computers distributed throughout the system would relay them from one "node" to the next in the general direction of their destination. If a packet failed to arrive, it could be resent automatically, without undue delay. Because the system was a spiderweb of redundant interconnections, node after node could be blasted out of existence, and the messages would simply be routed around the damage.

The first node of ARPANET—named for the Defense Department's Advanced Research Projects Agency, which funded the system's early development—went online at UCLA in 1969. The following year, MIT and Harvard joined the system. A year later, there were thirty computer systems in the Net. At first, access to ARPANET was limited to Pentagon-sponsored research facilities, but users soon turned it to their own interests. By the late 1970s, a community of science fiction fans was trading notes through the system. By the early 1980s, scientists and scholars around the world were clamoring to get into the Net. In 1983, the military set up its own MILNET for operational use and left ARPANET to the researchers. Now known as the Internet, the system was operated for several years by the National Science Foundation. Since 1987, the National Science Foundation's NSFNET has been managed by a corporation called ANS, a nonprofit collaboration between IBM and MCI.

Over the years, the system has grown dramatically and been upgraded several times. By early 1994, there were an estimated 2 million computers linked to the system and somewhere between 20 million and 30 million users, and the population was growing at an estimated 15 percent per month. No one really knows how many have joined the Net since then, and today's estimate will be hopelessly obsolete by next week. At this point, the Internet's main "data highways" can transmit roughly 5,000 book pages of information every second. After the next upgrade, it should be capable of delivering the entire contents of the Library of Congress several times each minute.

In 1991, the process got a sudden boost from the High Performance Computing Act, spearheaded by then-senator Albert Gore. It authorized the spending of more than $1 billion to upgrade Internet into NREN, the National Research and Education Network, designed to link schools, rural doctors, and just about everyone else into a nationwide data net. Exactly what form that network will take and how it will be administered are still being hammered out. One thing is certain: Commercial users were banned from the Internet until early 1994; they are welcome to do business on the new, improved Net.

WELCOME TO THE NET

While writing an article for a trade journal recently, one of our colleagues needed technical details about several new computer chips. A year ago, it would have meant bundling up for a two-hour round trip through the icy New Hampshire winter to the nearest college library, an hour or two of poring over printed indices several months out of date (this library is not as computerized as some), and long waits while the librarians searched the stacks for journals containing whatever articles he had found. Instead, he spent ten minutes searching a database of computer publications available through CompuServe and located more than forty references to promising articles. Ten minutes more, and he had five full-text printouts containing all the facts he required.

A short while later, a friend's wife was diagnosed as suffering from morphea, a mysterious and threatening disease of the skin and connective tissue. Ten minutes of research in Medline brought reassuring details that her physician had not thought to provide. Again, the alternative would have been a long, inconvenient trek, and he probably would have wound up settling for less information than he found online.

And while doing the research for this book, we needed to interview a physicist who has made a specialty of new mass-transit technology. To our dismay, he had just left for two months at the American research facility in Antarctica. (You have our word on this, folks! We know that it sounds like we are exaggerating for effect, but this story is true.) Fortunately, though we are new to life online, we remembered to ask whether he had an e-mail address in Little America. He had, and our query went out immediately. It was the following day before he checked for messages, but his reply still arrived within twenty-four hours.

To date, the information superhighway remains only a few steps above an information cart path. Yet it is already changing for the better the lives of many ordinary Americans. Like us, our friends are relatively new immigrants to the land beyond the computer screen. Our uses of its resources remains fairly prosaic. Even today, far more is possible.

In a book called *The Virtual Community* (Addison-Wesley, 1993), writer Howard Rheingold recounts his experiences with the WELL— the Whole Earth 'Lectronic Link—a pioneering online community based in the San Francisco area and stretching across the world. A spin-off from the old *Whole Earth Catalog,* the WELL combines access to remote databases with electronic messaging and special-interest conferences on subjects ranging from the UNIX computer operating system to the Grateful Dead and from gay sex to child rearing. Members meet, form friendships, have arguments, collaborate on work, play video games, shop, and gain access to information not available in any one of the world's largest libraries. Their bonds often are as close as any formed here in the outside world.

Rheingold tells the story of one "WELLite," known and liked in the virtual community, who left for an extended stay in the Himalayas. Nearly a year later, she came down with severe hepatitis and was hospitalized in New Delhi. Within hours after hearing of her plight, her online friends had located liver specialists in India, sources of rare medical equipment, schedules and prices for medical evacuation, and all the other details it seemed would be needed to keep her alive. They had also started a fund to pay her expenses.

Such communities are no longer uncommon. Another colleague of ours worries about his teenage son, a consummate "Net surfer." Like a surprising number of technologically savvy people his age, the boy may have more friends whom he knows only online than he does in his immediate community. He keeps in touch with them through e-mail and in piggybacked conference calls with up to twenty-five participants, simultaneously holding private side conversations over his computer.

None of this disturbs our colleague, not even the telephone bills. What bothers him is that his son uses the Net to keep in touch with an old private-school roommate, sending him articles clipped from popular American magazines and digitized into the computer with a scanner. The friend is the scion of a politically prominent Iranian family, most of his relatives are rabidly hostile to the United States, and much of the material our colleague's son sends over the Net is contraband in Iran!

Almost anything you want to do this side of cross-country skiing already can be accomplished faster and more efficiently over the Net. And it's getting better all the time.

More than ten years ago, an acquaintance who then ran a small computer company in Ithaca, New York, found that the cheapest way to communicate with customers in West Germany was to send them e-mail through The Source, a commercial bulletin board system since absorbed by its competitors.

Not long ago, the Millennium Project Feasibility Study at the United Nations University, in Washington, D.C., looked into the cost and practicality of sending questionnaires to some 200 study participants. Airmail would have cost about $1,000 per round of questions, and the trip to and from China would have taken a month or more, if the mail got through at all. Faxing the questionnaires would have taken only a few hours, but the bill would have amounted to $4,000. E-mail was all but instantaneous, and the cost came to a grand total of $20!

And in January 1996, another friend wanted to buy a new car. He chose the model with the aid of friends he had met online. Then he ordered through a computerized brokerage called Auto-By-Tel, which acts as a sales agent for some 1,200 auto dealers around the country. A week or so later, he picked his car up from a local dealership. There was no haggling, no high-pressure salesmanship. And because the service cuts the dealer's marketing costs from about $400 per car, on average, to $30, the price was the best available. No fewer than 12,000 people checked in with Auto-By-Tel that month.

By now, a few technophobes leafing through these pages in a book-store are probably saying "This is all very interesting, but what do these computer nerds have to do with me?" At the moment, perhaps nothing. But over the next five to ten years, the information superhighway will become much easier to travel, it will take on a broad range of consumer services, and the operators will build an entrance ramp right in your office and living room. Today, online information services are a conve-nience that many people find more than repay the modest effort and cost of learning to use them. Tomorrow, they will be easier to use, no more costly than cable television—which they may have replaced—and an unavoidable part of both work and play.

Those services probably will be pretty prosaic at first. Despite all talk of the World Wide Web, with its refined graphics and fast-growing cus-tomer base, by far the most popular use of the Net is still simple e-mail. According to the Electronic Messaging Association, a trade organization for companies that supply e-mail technology, between 30 million and 50 million people now use this service, almost 20 million of them in North America alone. Something over 6 billion e-mail messages each year travel over the world's data links, and the electronic clamor grows daily. In just ten years, e-mail traffic will total more than 30 billion messages annually.

One reason is that Web sites remain largely a consumer attraction, while e-mail is a profitable business tool. And as rapidly as individuals are signing up for Net service, companies are going online even faster. Many executives grew accustomed to basic e-mail service when their companies installed in-house nets in the 1980s and early 1990s, and now that global e-mail is available, they want it. A recent survey of *Fortune* 500 companies found that executives at all 500, every one, now use e-mail to communicate quickly and cheaply with colleagues who may work for divisions half a world away. Second-tier firms are scrambling to catch up. In the process, they are helping e-mail to grow as rapidly now as fax systems did ten years ago.

The other reason e-mail dominates the Net is that it is a remarkably effective way for people to exchange information. And information, rather than pretty pictures, is what many Net users want. All those conferences on the WELL and their counterparts at America Online, CompuServe, and the Internet operate by simple e-mail. So do thousands of mailing lists, which carry messages between readers with common interests that range from fly fishing to the chemistry of buckminsterfullerenes. Despite all the more sophisticated activities available on the Web, e-mail will remain the largest and fastest growing of the Net's applications.

Michael Dertouzos, director of MIT's computer science laboratory, cites three more services that seem likely to grow quickly in the networked world. They each have proven markets, and all of them are at least as mundane as e-mail. France's Minitel computer network, Dertouzos points out, has 5 million users and offers 15,000 services. Yet with all that diversity, just three functions account for the great bulk of Minitel transactions. If you need a phone number, Minitel makes a great alternative to fumbling with printed directories; that is the system's number-one use, as an electronic Yellow Pages. In second place, people with freight to ship use Minitel to find shippers. And the French buy their tickets over Minitel—tickets for movies and sporting events, even train tickets. This list can be humbling for those with grand visions of a networked world.

Experience in the United States has been remarkably similar. As of 1994, the four most popular services on the Prodigy system were news and weather reports, stock market information, and ordinary e-mail. Computerized shopping came in a distant fifth. And now that the online services all offer Internet access, e-mail is becoming a much larger part of their message traffic.

Of course, other uses are likely to grow as well. Raymond Smith, CEO of Bell Atlantic, which is making a major push into Net-based services, predicts that just five of the Net's thousands of possible applications will pay the way for the rest.

- Home sales: The Home Shopping Network and its competitors were instant multibillion-dollar successes when they reached cable television. The Net will soon bring interactive sales. Forget about calling the 800 number at the bottom of the screen. Just type in your credit card information—soon you will be able to slip the electronic equivalent of a credit card into a slot in your "information appliance"—and buy whatever appeals to you, either from televisionlike shop-at-home programs or from hundreds of online catalogs.

- Games: Fanatical fans around the world are already playing multiuser Dungeons and Dragons games over Internet. And AT&T is backing the ImagiNation Network, a nationwide system dedicated especially to interactive games. Using videogame equipment from Sega Genesis, 3DO, and AT&T itself, players in New York will be able to compete against players in Texas or Oregon without ever leaving their living rooms.

- Gaming: Once the network is wired, there is no reason not to hook in state lottery computers and deliver the online equivalent of the "instant-win" tickets now found in newsstands and tobacco shops.

- Video on demand: Although several interactive television experiments have flopped ignominiously, most industry analysts are betting that this will be the biggest moneymaker of all. Just glance at the online catalog, punch in the right number, and you can be instantly watching almost any movie ever made—and stop it, rewind, or fast-forward at will. For several years now, Tele-Communications, Inc., in Denver has been running a primitive prototype of the system. When viewers call in, clerks run to vast racks of videotape and pop the desired movie into a VCR. Despite delays of up to five minutes, customers reportedly have taken to the idea. And in West Hartford, Connecticut, Southern New England Telephone has been experimenting with a similar system.

It has proved popular enough that the phone company soon will open its own cable television service—though not yet video on demand—throughout the state.

♦ Online education: As early as 1994, Bell South announced plans to give 26,000 public schools access to its data network. It was the first step in a process that eventually will link every classroom in the country to online lectures, tutorial software, the Library of Congress, and no doubt educational services that have yet to be imagined. They will be accessible at home as well, to anyone linked to the global information network.

However, this is just the beginning. The profits from these and other ho-hum Net services will pay the way for more imaginative uses. All the contents of all the world's libraries; all the entertainment resources of Hollywood, Paris, London, and Rome; all the lecturers at Harvard and Oxford, MIT and the Sorbonne; and a good deal more are coming soon to a living room, den, kitchen, and classroom near you.

NET WORK

If the Net offers convenience and entertainment for consumers, it is also changing the way companies do business. In the high-pressure environment of the global marketplace, the vendor who offers the best prices, fastest deliveries, and the most convenient service can win customers all over the world. Increasingly, the company that makes the best use of information technology dominates its competition.

Advertisers already have mastered that lesson. At this point, no one can watch prime-time television without seeing commercials that direct them to a site on the World Wide Web. CNN has its Web site, where viewers presumably can find information more detailed and up-to-the-minute even than the stories that appear on its cable TV service. So do many local television news programs. Advertisements for movies aimed at the under-thirty crowd almost invariably offer a Web site for related material. Pharmaceutical advertising offers detailed product information at the firm's Web site. Even daytime television programs offer Web addresses where devoted viewers can learn more about characters, actors, and events associated with them. Our favorite (which we have never visited but whose existence we learned of while flipping past TNT for a midafternoon news update from CNN) is a Web page for fans of "The Wild, Wild West." There is something fascinating in the idea of a site on the World Wide Web dedicated to a television series whose original run ended before the Internet was born and nearly a decade before the first personal computers appeared. Clearly, as a force in popular culture the Internet has "arrived."

It is making itself felt in more useful ways as well. You probably have seen the Federal Express commercials extolling the firm's package-tracking system. Just have your computer call their computer, and you can find out instantly where your delivery got to. Some 10,000 people do so every day. Industry analysts credit that sophisticated information service with giving FedEx a commanding lead in the overnight shipping business. It works via the Internet and is just one of many such services that will appear on the Net in the coming decade.

Even manufacturers, traditionally slow to adopt new technology, are finding their way to the Net. One reason is a concept called JIT. Japanese companies introduced "just-in-time" manufacturing a decade or so ago. In the past, companies ordered raw materials and parts for their products, stored them until they could be processed and assembled, and then warehoused their merchandise until the marketing department could sell and ship it. That ties money up in inventory, causes delays when materials fail to arrive on time, and leaves backlogs of decaying merchandise when someone miscalculates the market. JIT pulls material through the factory according to demand. Products are made only to replace merchandise sold from the minimal stock on hand. When the final assembly area runs low on a part, it notifies the previous step in the assembly line that more are needed. That stage produces its parts only when the final assembly area needs them, calling back up the chain for its own materials as they run short. And so on. In this way, the capital tied up in inventory is kept to a minimum. The same process works well for retailers. Order exactly what you need when you need it, and you make the best use of your money.

Making JIT work depends on two things: suppliers who can deliver the goods almost instantly when they receive an order and a reliable flow of information at all stages in the manufacturing process. Many manufacturers now refuse to deal with suppliers who cannot make deliveries daily; some require shipments twice each day.

The process grows even more efficient when you network your company's computers with those of your supplier. That way, there are no delays for order processing. When your clerks enter an order, it appears instantly on the supplier's warehouse or factory floor computer. A growing number of companies now buy only from suppliers who can work this way; this is one reason Wal-Mart can keep its prices low. Some firms are even linking to the computers in their suppliers' manufacturing equipment. Enter an order for goods, and the supplier's production line automatically begins to turn out the items you need. This is clearly a job for e-mail; already, most of those orders flow across the Net.

This does not just cut inventory costs. Better information flow translates to better customer service. In an interview with *Wired* magazine, Apple founder Steve Jobs points out that it can take three months to special-order a car. The reason is that information flows through the manufacturer's procurement system at glacial speeds. The car itself could have been built in a week, but it takes months just to order the pink paint and purple leather for your new Cadillac from suppliers who already have it sitting on warehouse shelves. Streamline the ordering system, and every car could be tailored to meet the buyer's whim. That will make for happier customers and fatter bottom lines.

In ten years, all but the most backward companies will be able to route their orders and shipping data over the Net, both to meet the demands of JIT and to customize products for the end user. The rare holdouts will be courting Chapter 11. And we consumers will be able to order, for immediate delivery, products with exactly the combination of style, color, and features that best fits our taste and practical needs.

For workers, the Net giveth and the Net taketh away. Unfortunately, many of us are finding that it takes away before it gives anything in return. It used to be that a corporate manager could effectively oversee only six subordinates. With computers, and especially given the fast flow of information made possible by the Net, one manager can watch over up to twenty-one subordinates. This, and not the low wages paid to foreign workers, is why many middle-class Americans now feel their jobs are on the line. In the 1970s and 1980s, business found that machines could do the work of people on the assembly line. More recently, computers have been replacing managers. Companies simply do not need as many human beings to get the work done. The jobs now being lost to computers and the Net will not be replaced until new employers come along to create new openings.

Chances are that those new jobs will involve the Net. At the end of this decade, about 45 percent of the American labor force will be working in the information industries. By 2000 or so, more than 80 percent of all managers will spend their time collecting, analyzing, synthesizing, structuring, or retrieving information that arrives and is passed on over the Net.

For American workers, this transition has brought difficult times. Yet in the end, the Net represents their best hope for prosperity. Information is one of the few areas in which the United States has a significant lead over its competitors. Some 62 percent of American households subscribed to cable television as of late 1993, compared with only 2.7 percent in Japan. American companies had compiled 3,900 domestic commercial databases, Japanese firms only 900. American firms had installed 417 personal computers per thousand workers, Japanese companies only 99. This is one reason the American workforce has until recently found its pay scale declining faster than that of Japanese workers. It is also why American working conditions have a better chance of recovery, at least for the computer literate.

The gap is even wider when it comes to the Net. As of late 1993, there were 1.18 million computers connected directly to Internet in the United States, only 39,000 in Japan. At this point, the number in the United States probably has doubled, according to the best estimates we can find; in Japan, it has barely changed. No one has a comprehensive list of the Internet access providers in the United States, but most colleges supply connections for their faculty and students, many corporations routinely tie employees into the Net, and commercial Internet services clearly number in the hundreds. As of mid-1995, there were ten commercial access providers in Hong Kong alone. Japan has only one, and that one is owned by Americans and supplies service almost exclusively to Japanese branches of American companies. If the United States handles this opportunity well, its dominance of the Net should give American companies a commanding lead in the global market for information and services. This will give highly paid American manufacturing workers a competitive advantage over their low-wage competition elsewhere. It is the only hope we see to stem the decline of real wages in the United States.

OUR LINKS TO THE WORLD

Technologically, only three questions will be important in the next few years: How will all this information reach our homes and offices? Will we read (and watch and listen to) it as we do today, on a personal computer, or will we use some other device? And can the Net be held to software standards, so that all "Netizens" can use the same access programs to visit all sites? A long-run challenge is how to expand network capacity to carry all the data the customers will want.

For the first two questions, the shortest answer is probably the best:

Until competition weeds out the weakest suppliers, everyone who can deliver data will; whatever hardware can be adapted to receive it will be. Visions of profit are dancing in the heads of executives throughout industry, and no one wants to be left out. The result has been a stampede such as no one has seen since a mill owner named Sutter found gold in a California creek almost 150 years ago. Cable TV operators, telephone companies, and electric companies are some of the obvious competitors.

Cable TV operators may be closest to owning a piece of the information superhighway. They already pump one channel of video data into two-thirds of the homes in the United States. Replacing their coaxial cables with fiber optic lines would let them carry all the information required by the Network Age. Continental Cablevision and several smaller cable providers have already begun linking their customers to the Internet. Many more will follow.

The telephone companies are almost as well positioned. In California, Pacific Telesis has already announced plans to replace the copper telephone lines serving almost half of the firm's 10.4 million client households with fiber optic lines capable of carrying far more information. About 5 million homes should be linked to the system by 2000. Two other regional telephone companies, US West and Bell Atlantic, are building similar data networks. Bell Atlantic and US West have even contemplated deals with cable TV operators outside their regions to supply telephone, interactive television, and data service.

Both cable and telephone systems have technological hurdles to get over before they deliver practical Net service. For telephone companies, the problem is data speed. The fastest modems found in any Wal-Mart send and receive data about as quickly as ordinary phone lines can handle. To carry video, telephone companies must switch to fiber optic lines, at least to carry their signals into each neighborhood. Eventually, the wires leading into each house will also have to be replaced. Cable operators face a similar problem. Their older lines provide only a one-way data path, and even the newer two-way cables were intended only to let them query the set-top tuner about its status. Cable companies also have some new lines to string. Whoever gets the job done fastest probably will capture the Net service market.

A third, less obvious, group of candidates also is seeking to become Net providers. It turns out that many of the nation's electric companies already have their own two-way fiber optic networks, leading almost to the consumer's door. Used originally to keep the utility headquarters in contact with substations, the power companies plan to expand the systems so they can read each home's electric meter without anyone ever leaving the office. Greater efficiency alone should repay their investment. But there is more to come. Baltimore Gas and Electric has been offering its fiber optic network to business customers as an alternative to the local telephone lines. Entergy Corp., of New Orleans, has already wired fifty of its customers in Little Rock, Arkansas, into its network and plans to tie in 440,000 homes by the end of the 1990s. It needs only 2 percent of the system's data capacity for its own use; the rest will be leased to data and entertainment services.

Add a host of wireless communications operators to the list. At least five companies and corporate alliances plan to launch satellite-based communications systems:

- Motorola is scheduled to launch the first of sixty-six low-orbit satellites that will form a global telephone system known as Iridium. By 1998, Iridium should link personal computers, pocket telephones, fax machines, and pagers in a global network.
- Teledisc has even larger plans. A $9 billion project, it aims to send no fewer than 840 satellites into low-Earth orbit. The firm calls its system an "Internet in the sky." By 2001, if all goes according to schedule, Teledisc will provide global data service at speeds up to 2 million bits per second, about seventy times faster than the fastest modems now linking home computers to the Net. Though skeptics view this project as too ambitious to succeed, backing by Microsoft founder Bill Gates and other luminaries of the high-tech world mean it must be taken seriously.

- Nextel Communications is planning to extend specialized mobile radio, now used largely by truck companies and radio-dispatch taxis, to serve as many as 180 million customers in twenty-one states and nearly all of the fifty largest cities.
- Bell South, Cox Enterprises, MCI, and other companies are also developing specialized personal communications networks, an advanced portable telephone system expected to be cheaper and more versatile than conventional cellular phones.
- Most recently, ICO Global Communications has announced plans to clutter space with still more satellites. Its system will occupy "intermediate circular" orbits and provide data, fax, phone, and paging services. This project is backed by Inmarsat, a consortium of seventy-nine member countries that already operate communications satellites.

Though only the Teledisc system will be optimized for data communications, any of these competing systems could be linked to the Net. All of them almost certainly will be.

For Net users in the developed world, wireless data systems represent a convenience that will keep us in touch with the world's data stream wherever we go. In many other regions, they will bring both telephone and data service to people who have never even heard a dial tone. As early as 1994, Argentina licensed a consortium called CTI to install wireless telephone systems throughout the country, much of which lies far beyond the copper cables that serve Río and other major cities. The first system was up and running just four months later. Thailand already is becoming the first country to bypass copper wires entirely. Its telephone system, formerly limited to a few major population centers, is now building a nationwide cellular phone network that will link even the remotest corners of the land to the outside world. In future, the technological and economic backwaters of Asia and Africa will join the modern world by satellite telephone—and that is all it will take, a pocket telephone talking directly to a relay satellite. Net service will follow immediately.

In the end, it will hardly matter who brings us our data or carries our messages out to the rest of the online world. Whether those optical fibers are owned by a company that started out in cable television in the 1980s or by a utility with a century of telephone service behind it, the data remains the same. In the United States, the two groups will coexist. Early in 1996, in the name of promoting competition, Congress moved to deregulate the communications industry; in the process, it freed phone and cable companies to avoid competing with each other. Rather than waging a war to the death, they will cut deals to share the market for data service. In the end, most homes and offices will be served by a single data path, which will carry the digital equivalent of telephone, television, e-mail, and all the other data services on which the developed world is coming to depend. By 2005 or so, nearly everyone in the United States, Canada, Western Europe, and the developed regions of the Far East will have access to the Net. So will anyone in the Third World who is willing to pay for satellite access.

The other half of the hardware is the equipment we use to tie our homes and offices into the Net. Here there also will be many choices.

In one picture of the future, the principle of convergence works its inanimate will. The old PC, if we had one, probably will be gathering dust in some remote corner of an attic, along with the owner's television, telephone, and most of the home electronics that many of us now take for granted. What remains will be a one-unit-does-all data communications system, something like a super-high-definition television with a box on top that resembles today's cable TV tuners. Somewhere nearby will be a keyboard, for those who still feel the need to type, and a headset for those who do not; the headset will also replace the telephone for conversations with other people. This is the "information appliance" that we first heard of when political fashion had Vice President Al Gore proclaiming the wonders of the "National Information Infrastructure." Cable TV companies are rooting hard for this version of the information superhighway, because they think it puts them in an ideal position to collect the tolls.

In another version of Net's evolution, our digital, high-definition television-for-the-twenty-first-century will remain largely an entertainment device. We probably will keep a separate telephone as well, simply because it is convenient to segregate some functions into special-purpose equipment. But for serious data management—and that includes communicating across the Net—the personal computer will still be king. All these devices—computer, telephone, and TV—may be connected to the Net through a so-called information furnace, an interface device just bright enough to recognize where all those incoming bits belong and route them to the telephone, television, or computer as needed. AT&T and its local competitors love this idea, because they imagine that it could help them retain a hold on data delivery. Rather than paying for that information furnace all at once, they hope that customers will prefer to rent it from their old, familiar phone company.

There is a third option as well: Forget about expensive home computers and information appliances. Keep the brainpower on the Net itself. Then customers can use an inexpensive "Network computer" to log in to a Net-based computer services. This notion has at least two advantages: When a new version of your word processor comes out, the service provider adopts it automatically; customers never have to buy an expensive upgrade. And the machine itself is cheap, costing a few hundred dollars at most for a device that would use your television as its display screen. Almost everyone should be able to afford one of these stripped-down network terminals, even if they use the Net only casually.

However, these terminals also bring their own disadvantages. Your files, like the system's intelligence, all are stored on the service provider's computers rather than in your own home or office. That brings up some serious questions about privacy. And when the provider's customer base outgrows its computer system, service could get s.l..o . . . w. Or it just might die altogether. When was the last time you tried to use your bank's ATM, only to discover that "the computer is down"? This is the *Back to the Future* version of the Net, in which we all return to the days when computer users sat at a dumb terminal waiting for a distant mainframe to respond—petitioners at the feet of the oracle.

To many hard-core computer users accustomed to salivating over the latest, most powerful PCs, this idea seems as dumb as the terminals themselves. Yet a few computer-industry leaders are very much taken with it. They see it as their best chance to wrest market dominance away from Microsoft, whose Windows operating systems now run in eight out of every ten personal computers in the world.

Which of these competing visions will provide tomorrow's Net hardware? For the near term, the answer is obviously the PC. In the long run, probably bits of each.

For at least the next few years, technophiles will continue to dominate the consumer market for Net access. That effectively closes the door for dedicated Net terminals. Hard-core computer users will stick to their PCs, and people whose eyes glaze over at the first mention of megabytes and bits per second will be as repelled by so-called Network computers as they are by the real thing. In two or three years, and perhaps much sooner, the notion of resurrecting the dumb terminal as our primary link to the Net will be put to rest.

The first practical "information appliances"—call them Net televisions—will not reach the market until 2000 or so, and they will not become really useful until digital TV arrives a year or two later. However, by 2005 Net televisions will have absorbed that hypothetical "information furnace." With a little luck, the result will be more useful than the entertainment and home-shopping systems that many computer-industry executives now envision.

Picture the networked home of the 1990s. At one corner of the house, a single broad-band data path enters. On the way in, it carries e-mail; digital telephone service; digital television, both scheduled programs and video on demand; billing queries from the electric company and other utilities; and whatever other information services have arisen in a fertile marketplace. On the way out, it carries e-mail, requests for movies, billing information, and so on. The data cable is attached to the Net television, for no better reason than that this is where consumers are used to attaching their cable.

That costly information furnace the phone company once hoped to rent you has turned out to be a single chip inside the Net TV. It costs less than $20 and routes data along the home's internal network. Video signals may be held back for the TV's own use or passed to monitors throughout the house. E-mail goes to whichever computer is in use or, if several are on at once, is divided appropriately among them. Some of it may even go to a dumb terminal. At the same time, a secondary controller configured for the household takes care of all the internal communication between the home systems—security alarms, lighting, lawn sprinkler, heat and air-conditioning controls, and so on. This is where the billing messages from the power company go and where the responses originate. If need be, the home messaging system can even relay it to our "personal digital assistant"—a glorified pager, for all that hardware companies would like to see it as a revolutionary advance in computing. It is a complex mechanism but almost invisible to the user. All we know, or care about, is that the information we want can find us, wherever we are.

Software already is presenting much more difficult problems. What made the Internet work is a common set of protocols, programming standards for the transfer of information that could be adapted easily for any computer. Thus it does not matter whether we use the Internet from an IBM–compatible PC, an Apple PowerMac, a UNIX workstation, or a dumb terminal attached to a shared mainframe. Given software that adheres to the standard, everyone can travel the world on the Net.

When the World Wide Web came along, it carefully maintained this compatibility. Web pages can display both text and graphics and "link" to other Web pages by virtue of a standard called HyperText Markup Language (HTML). (Web pages, for those who have yet to create one, are just text files peppered with HTML codes that tell Web browsers how to display them and where to find linked documents.) Without that standard, there could be no World Wide Web.

Suddenly new would-be standards are pouring out of Net-minded companies and research centers. Some merely improve on HTML. Others offer powers new to the Net. Netscape Navigator brings more sophisticated layouts to the Web; it provides the only way thus far to display tables on a Web page. Adobe Acrobat, another formatting tool, is said to give Web sites the look of a printed page. Microsoft's Internet Explorer has added an audio standard to its browser, so Web pages can include music and speech. Just over the horizon, a Microsoft project codenamed Blackbird reportedly will pack online documents with better graphics, audio, video, and even animation.

At least three competing standards offer to bring FM-quality sound to Web pages. One—RealAudio, from Progressive Networks, in Seattle—has already been used to build the first Internet-only "radio" station. Just "http:" to the Web site for Radio HK, and music or commentary pours from your computer's loudspeakers.

Virtual Reality Modeling Language (VRML) offers even greater novelties. With VRML, objects on a Web page will move realistically through their own three-dimensional landscape. Stunning new Net-based video games are a given. Look forward as well to graphical search engines that enable you to walk through rooms or fly among buildings, each one a data provider somewhere on the Web. In a few years, the artificial world of science fiction cyberspace will take form on the Net.

By far the most revolutionary new standard is Java, from Sun Microsystems. It can deliver small applications programs to your computer along with each Web page. A page of technical data, say, may soon arrive with its own statistical package to help make sense of the raw numbers. The program, called an "applet," vanishes when you are done and is downloaded again whenever you call up that Web page. In effect, the document itself has grown a brain. Industry pundits see in this a whole new way of computing, with the Net as its heart and personal computers as little more than appendages.

The problem with all these standards is that they are not really standard. They are competitors, and a program that works with one by definition cannot work with the rest. Netscape Navigator has become the de facto standard for creating more sophisticated Web pages than are possible in HTML, and Java has won the crown in the nascent applet field; even Microsoft has swallowed its corporate pride and bought into the Java standard. But where HTML was created and endorsed by the Internet's central operating authorities, Netscape and Java have nothing going for them but market acceptance.

In the end, that could be enough. By 2000 or so, only one relevant standard will survive—one method of presenting tables, one way to transmit audio data, one way to create virtual reality environments, probably even one way to write applets, all gradually evolving to greater power and versatility. Until then, we probably will find that more and more Web sites are closed to our Internet browser programs. Making that transition as quick and painless as possible is one of the greatest challenges the Net now faces.

That leaves one last technological hurdle, expanding the Net's capacity, so that it can carry all the data that will be demanded of it. Just one channel of digital video requires about 500 times as much capacity as a telephone call, or about 16 million bits per second (bps). At that rate, it would take about 18 minutes to download one frame of video from the Net to a home computer equipped with an average modem capable of receiving 14,400 bps. Some applications are even more daunting. For example, each frame of the movie *Toy Story* in its original all-digital form contains 300 megabytes of data, and the film consists of 114,000 frames. Downloading that film through a modem (before it was processed for video) would take about 600 years! An expensive, single-purpose data line could cut that down considerably, but the point is clear: Enough such services trying to make their way to enough customers could soon make the Net seem very crowded.

Anyone who has used the World Wide Web already has twiddled his or her thumbs while waiting for a page to appear. In part, this is because data transmission through regular telephone lines is relatively slow. However, many providers already are nearing the limits of their capacity. One user who calls up a Web page often has to wait until someone else's request is processed. This problem can only get worse as the demand for data grows to overwhelm not only the local provider's equipment but the major data arteries of the Net itself. The problem will become acute well before the end of the next decade.

One partial answer is data compression. If, say, a graphics image has a large patch of blue, there is no need to send 10,000 identical pixels of blue. Instead, send one pixel and the instruction to repeat it as often as needed. Compression can work wonders. Sending high-quality digital speech used to require transmission speeds of 32,000 bps. It can now be done at 13,000 bps. Reasonably clear speech can be sent at only 8,000 bps. Video and graphics data now can be compressed to one-fiftieth of its original volume, and sometimes less.

Eventually, however, we will be faced with the same problem—too much data, not enough carrying capacity. At that point, we will need new technology to replace our outmoded networks. Fortunately, it will be available.

Fiber optic cables already in use could theoretically carry as much as 25 trillion bits of data per second through each glass fiber. This is 100 times more than today's fastest networks achieve in practice. What slows the system down is the need to convert information from electronic form to pulses of light and then back again—electronics just works a lot more slowly than optics. Unfortunately, currently that change must be repeated every few miles, to amplify the data signal as it passes through the cable. If data could remain in optical form from the time it enters the Net until it reaches your computer or Net television, the Net could run at peak efficiency. And that would be fast enough to carry all the data anyone is likely to want in the foreseeable future.

Experimental all-optical networks have already been made. At AT&T's Bell Laboratories (now Lucent), scientists have packed up to

340 billion bits per second into a single optical fiber and transmitted it for distances of up to ninety miles. Researchers at the Japanese telecommunications firm NTT have demonstrated a packet-switched optical network capable of data speeds of 100 billion bps. This could well become a prototype for the next-generation Internet.

There is no guarantee that the all-optical networks now in the laboratory will take over the Net when the growing demand for data makes today's network obsolete. Electronic data systems remain far cheaper than pure optical equipment, and their performance is improving rapidly. And it is entirely possible that some unforeseen technology will come along in the next decade to replace even the most promising systems now in the works. What does seem certain is that superfast networks will be available before the Net outgrows the systems on which it now runs. For our purposes, that is all that matters.

POLICY QUESTIONS

We set out to write this book almost as if going on vacation. For once we could focus on the bright promise of technology and leave behind the tedium of economics and government regulations that often dominate our forecasts. The idea was as welcome as a Caribbean vacation in the middle of a winter filled with snow and sleet.

Unfortunately, the Net is one technological wonder whose character will be shaped almost exclusively by market forces, regulation, and even court decrees. In this one chapter, at least, we cannot avoid these issues.

Until recently, the Net grew if not exactly at random, then at least without much in the way of regulation. Though the National Science Foundation has paid about 10 percent of the cost of administering the Internet, individual nodes are supported by their owners, mostly universities and major research centers. This has allowed the Net a degree of freedom seldom found in the modern world. On many Net sites, almost anyone can use the facilities. "Anonymous" is all the password required to download information from many of the systems in the global data network; it often carries the right to upload files to them.

That is how most current Net users would like to keep it. Much of the value of the Internet grows from just this informality. The Harvard sociologist Daniel Bell forecast that in the post manufacturing economy, information would become a product, to be bought and sold like any other; to a remarkable degree, that prediction has come true. On the Net, in contrast, information is common property, maintained and augmented for the benefit of the community. Anyone who needs information probably can obtain it, so long as he or she can figure out where to look for it; anyone with something to contribute is more than welcome to do so. Many of the search utilities that now make locating information on the Internet somewhat easier were created and donated by Net users who had nothing to do with the Net's relatively informal "management," moved only by the wish to contribute to their chosen community.

For better or for worse—and many Net participants and onlookers believe it will be for the worse—this fertile anarchy may not long survive. The Net has grown too big and too important to escape government notice; hence the High Performance Computing/NREN Act (HPC/NREN) and the Clinton administration's plan to build the National Information Infrastructure. And government attention invariably means government regulation, as we have seen in the censorship provisions of the Telecommunications Reform Act of 1996. Further, the National Science Foundation has gradually withdrawn its funding from Internet. That support is to be replaced by private enterprise—the Net as advertising and sales medium. To business, information truly is one more salable commodity. And if something is to be sold, it cannot also be freely available. If tomorrow's data merchants are to make a profit, they must find some way to keep at least some information out of general circulation.

That brings up a number of important concerns. Who will be allowed to sell goods over the Net? What about intellectual property rights? Do

Net "publishers" retain the rights to their creations, or does material sent out over the system become public property? Who gets access to the Net, which after all was developed largely with public funding? Who, if anyone, will decide what may and may not be said on the Net? None of these policy questions has yet received a final answer.

One of them became urgent as early as the end of 1991, just a week after HPC/NREN became law. Since 1987, the National Science Foundation's NSFNET had been managed by a corporation called ANS, a nonprofit collaboration between IBM and MCI. Among ANS's powers, it can decide who gets access to NSFNET—in effect, who gets whatever profit is to be made now that the Net is open to private enterprise. Then ANS spun off a subsidiary called ANS CO+RE, to sell communications services on the Net. The plan would have allowed ANS to profit from its own commercial ventures while deciding who would be allowed to compete with them. At best, it seemed a conflict of interest, as the *New York Times* and others pointed out, and ANS quickly distanced itself from the scheme. So far, the free market rules, and whoever has a product to offer on the Net has been welcome to do so. Yet there is no guarantee that this will always be true. How to control access to the Net could be one of the most important policy questions of the next fifteen years.

What happens to intellectual property rights on the Net? The virtual community began as a land of dedicated researchers sharing scientific data and enthusiastic "Netheads," who tended to live by the anthropomorphic slogan "Information wants to be free." Neither group cared much about whether their material was copyrighted. But over the Net, print, pictures, music, and graphical material can all be scanned, copied, sampled, and distributed with or without alteration at will. That idea horrifies publishers and other media companies—not to mention some writers and artists—whose income depends on their power to control access to their works.

Though the debate over publisher's rights vs. free access continues, the key decision may already have been made. As part of a long-running personal war against the Church of Scientology, former member Dennis Erlich once posted some church documents onto the World Wide Web. Now a federal court in San Diego has ruled that doing so violated the church's copyright. The decision set an important precedent for the future of the Net, because it offered the broadest print-style protection yet accorded to material "published" on the Net. This is sure to encourage companies interested in selling proprietary information over the Internet. But it may spell the end of a freedom that many Net denizens have found precious and productive.

We do not foresee many lawsuits in which casual Net users are pitted against media giants trying to quell unauthorized use of copyrighted material. It is pointless to sue someone who illegally distributes someone else's intellectual property by computer; hardly anyone makes money from what they upload to the Net. Yet companies can now use the power of law to discourage unlimited use of information. In future, many segments of the Net could be off limits to all but subscribers, just as CompuServe, America Online, and the other independent commercial systems are today.

The other problem is who can use the services offered through the Net? The obvious answer is, whoever can afford to pay for it. But that is not quite good enough. Leaving aside the cost of the personal computer and modem needed for access to the system—good used computers these days go for as little as $300, and a high-speed modem can cost less than half that—computer communications is expensive. Hardcore computer junkies who cannot reach an Internet service provider with a local call often spend hundreds of dollars each month, just on telephone bills. As things stand, access to the Net often is limited to the well off.

Yet, the cost of Net service is not all that high when you consider what it offers. The average middle-class household that uses the telephone daily, takes a newspaper, buys a few magazines, and pays for one or two extra-cost cable channels already spends something on the order of $150 per month on information services. The Net will deliver much more at relatively little extra cost. But it will still be too expensive for the poor.

It is hard to argue that this is fair. As we have seen, computerized communications soon will offer major advantages to those who can use it. Net access does not mean access just to entertainment and home-shopping services but to education, information, and no doubt a host of business opportunities for the entrepreneurially minded. Info-starvation now appears to threaten the Net-deprived in ways that are deep and lasting.

The Network Age is not for the poor or the semiliterate. Even before the dimensions of the Network Revolution became clear, the Bureau of Labor Statistics estimated that during the 1990s, some 6 million jobs would open up in the technical, professional, and executive categories while scarcely 1 million would appear for unskilled workers. The growth of Net-based industries can only broaden this gap.

Some forecasters have seen in this the rise of a two-class society. On the economic high ground will be the well-educated, information-rich elite, who make a comfortable living in the challenging, creative industries now forming around, or being transformed by, the Net. The poor and the poorly educated, in this view, will live in poverty, their futures blighted by their inability to tap into the Net and use its information.

And, again, the Net was developed with public funding. Shouldn't everyone have access to it? Many people think so.

There are only three obvious ways to make certain that less affluent Americans get data service along with the rich. Make private enterprise pay for it as a condition of gaining approval to hook up more profitable neighborhoods. Have the government subsidize Net access for low-income groups. Or install public data terminals in libraries, community centers, and other such facilities, again at the taxpayer's expense.

Understandably, few companies are happy with the first alternative.

Bell Atlantic and Tele-Communications, Inc., for example, expect to spend $11 billion to rewire 8.75 million homes in the Washington, D.C., area before the decade is over. Even with that ambitious schedule, the project will leave 2.25 million homes in poor neighborhoods without connections to the Net. Forcing the firms to expand coverage throughout the region would cost them something like $2.75 billion more, which, in their view, seems unlikely to be repaid.

Alternatively, Washington could subsidize the construction. It would have to do so throughout the country, at a cost many times more than $2.75 billion. Because it does little good to provide services the poor cannot afford to use, add the cost of a continuing subsidy for the monthly fees and telephone bills that commercial data and entertainment services will bring with them. In a time of tight budgets, it seems unlikely that government will be willing to make such a commitment.

That leaves the third alternative, providing free-access terminals in public places. Several municipalities are doing so already, with considerable success.

Santa Monica's Public Electronic Network, or PEN, is open to personal computers throughout the area and can be entered through terminals in the city's schools, libraries, and other buildings. Users can read public announcements, join open conferences on a variety of subjects, and exchange e-mail with city officials. Accounts on the system are free to all; even some homeless residents are active participants. The system opened for service in 1989; by mid-1990, it had produced concrete results. Participants of one conference pointed out that the homeless cannot hope to find work without somewhere to wash, store their meager belongings, and clean their clothes. City government soon responded by installing lockers, public showers, and a free laundry for the homeless. A PEN terminal in a community center serves as a jobs bank.

Cleveland is another center of public networking. Its Free-Net system was opened in 1986 and soon attracted 7,000 users. Three years later, it expanded and added links to the network at Case Western Reserve University and to the Internet. Since 1987, similar systems have spread to Youngstown, Ohio, Cincinnati and Peoria, Illinois, and rural Medina County, Ohio. If the Free-Net idea continues to propagate, these systems could go a long way toward providing Net access to low-income Americans.

We have no forecast to make here. We do not know how this issue will be solved within the current restraints on government spending. We know with certainty only that an answer must be found quickly. The alternative is to deprive many citizens of a tool that will be increasingly critical in building a productive, satisfying life.

A third major question has serious implications for the right of free speech: What can be allowed on the Net? Until recently, the answer was anything at all. Again, that is just how most Net users like it. The Internet includes bulletin boards where members of the virtual community can leave messages for President Bill Clinton and Vice President Al Gore; discussion groups covering aquarium fish, antique autos, biotechnology, and rape; space photographs from the NASA collection; and a lot more. But many interest groups contain sexually explicit photographs and discussions as well as other material that people of tender sensibilities might find offensive. How much of it will survive the commercial world's least-common-denominator, let's-not-offend-anyone approach to merchandising worries a lot of people who care more about the First Amendment and their own access to whatever interests them than they do about business and propriety.

That does not include the U.S. Congress. The Telecommunications Reform Act of 1996 created heavy fines for the propagation of a wide variety of material that social conservatives find objectionable. Place a pornographic picture on the Net, and it could cost you as much as $200,000 if a child happens to see it. Send a message that includes any of the seven obscenities and otherwise impolite words traditionally forbidden to broadcast media, and that too could cost you. Predictably, discussing abortion is one more offense subjected to costly fines. One federal court quickly banned enforcement of some of the act's more draconian limitations on free speech, but its opinion was so ambiguous that partisans on both sides of the censorship debate were left wondering whether they had won or lost the argument.

A CAUTIONARY LOOK AHEAD

Civic boosters call Blacksburg, Virginia, "Exit 1 on the Information Superhighway." We think of it as a guidepost en route to the Network Age. A community of about 34,500 people nestled in the foothills of the Alleghenies, the people of Blacksburg have found one solution to the problem of universal Network access. There are lessons here for everyone concerned with life in the wired-up world of the early twenty-first century.

Aided by Virginia Tech and Bell Atlantic, city fathers have wired up the town, a project they call the Blacksburg Electronic Village (BEV). Since the project began some three years ago, the town's public library has trained more than 20,000 people in how to use the Internet. About one-third of the town's residents and 145 local businesses have signed up for e-mail service, unlimited access to the global Internet, and a host of local bulletin boards. This service costs them all of $10 per month. BEV even has its own home page on the World Wide Webb: http://www.bev.net.

To date, e-mail seems to be the most popular Net service among Blacksburg residents, as it is throughout the Net-using world; they send roughly 250,000 messages every day. Already, one local market, two bookstores, a clothier, and even a jewelry store accept orders over the Net. Soon town water and sewer bills, and even parking tickets, will be paid by computer. Local organizations such as the Girl Scouts and the League of Women Voters report that attendance soars at meetings that are announced on the Net. Several of the businesses that opened Web sites to attract local customers report that they have received inquiries from potential buyers from distant parts of the globe.

Thus far, BEV is by far the most advanced community-based Net program, but other towns are quickly following Blacksburg's lead. Among them are such diverse communities as Santa Monica, one of the classic bastions of wealthy California liberalism; Denver, Colorado, a metropolis rife with inner-city problems; and Ridgewood, New Jersey, an affluent New York suburb. BEV director Andrew Cohill estimates that it would cost most communities about $300,000 to set up the technological base for an "electronic village" and about $100,000 per year to operate the system. This is an investment that many more towns and cities will make over the next decade.

Blacksburg's experiment offers a glimpse of one possible future, not just of community life but of our personal lives, both at work and at home. Some thirty years ago, media theorist Marshall McLuhan forecast the coming of the "global village," a planetary culture unified by the flow of information. People across the world, he believed, would come to hold common values and opinions because they marinated in the same broth of information, which was filtered through the same mass media. Instead, if we follow Blacksburg's precedent, we will form millions of "Network villages," each with global access. We will be united by the Net but distinguished from each other by the information filters we create for ourselves.

Network villages are not merely information consumers but information producers. Just as Blacksburg has its own home page on the World Wide Web, so will a billion other villages across the planet. Some will be physical communities like Blacksburg. Many others will be intellectual communities, with individual members scattered across the planet but united by an interest in particle physics, new treatments for schistosomiasis, the music of Enigma, or the restoration of early Harley-Davidson motorcycles. Each will generate its own information and contribute its messages to the rich stew of data that fills the global Net.

However, other futures also are possible. The high-speed data lines soon to arrive at our doors will be built by private industry. They will be installed primarily to sell us their creators' products without having to pay a middleman to distribute them. These will be largely one-way services; tell the video company which movie you want to see, and it will deliver what you have ordered. If these commercial services dominate the Net, it could become an information superhighway into our living rooms but an information footpath in the other direction. Even in Blacksburg personal e-mail and interactive hobbyist groups could persist only as an afterthought.

MIT's Michael Dertouzos imagines the Net as being much like society at large. We will use it, he suggests, for business and personal communications, to market products, to move X rays from the lab to the doctor's office, to get an education, in roughly the same proportions society gives these purposes today. Business inevitably will occupy much of the Net, because that is how we spend our time and energy here in what Net habitués refer to as "the real world." We suspect he is right.

Yet this is an important issue, and it is receiving very little attention. The next ten years will fix the character of the Net, probably for decades to come. The decisions that will establish its nature are matters of public policy. We would like to see them made by informed debate among the millions of people who will have to live with the result. We fear, however, that they will be made by default.

BRICKS FOR THE
HIGH-TECH FUTURE

What do magnetically levitated trains, the next generation of space launchers, and advanced microchips all have in common? None of them can be built from the steel and aluminum, plastics and silicon on which everyday technology depends. Instead, dramatic advances in these and many other fields require advances in materials science, the hidden technology that underlies the more obvious miracles we usually think of as the ultimate in "high tech."

Almost nothing in today's world could exist—certainly not in its current form—without materials that were no more than laboratory curiosities a century ago, if they were available at all. Skyscrapers were born when steel was state-of-the-art technology; but today's giant buildings are sheathed in glass layered with heat-deflecting coatings, and they are laced with plastic-coated wires and data cables made from fibers of glass purer than anything known just a few decades ago. Cars are still made largely of steel much like that from which their ancestors were wrought; but their engines increasingly are built of aluminum alloy (aluminum itself was not commercially available until the 1890s); their wires are insulated by plastic rather than cotton and paper; and their operations are controlled by chips of silicon, which could not be made pure enough for electronics until the 1960s. Carpets have been woven for at least 4,500 years; but the nylon from which many of them are made today was discovered only in the late 1920s. From the insulation in your walls to the heat-proof bowls in your microwave oven, most of the conveniences we take for granted could not exist without materials that had not been dreamed of a century ago.

One hundred years from now, historians of technology almost surely will say much the same thing. Many of the materials they will use late in the next century will be things that we cannot even dream of now.

In fact, just ten or twenty years from now, engineers will rely on innovative new substances that today are just finding their way out of research laboratories. Some will be closely related to the materials we already know. These include new high-performance alloys, plastics and composites that can survive extreme temperatures, ceramics that offer greater strength and shock resistance. (The first ceramic auto engines will reach the market no later than 2005.) And wholly new classes of material will soon be available.

High-temperature superconductors will finally reach the market in the next decade. The first will operate in baths of readily available liquid nitrogen; room-temperature superconductors should arrive—this forecast comes without a guarantee—between ten and twenty years later. At first, cost will limit superconductors to special uses, but by 2020 they will enter some consumer products.

So-called intelligent materials can monitor their own condition, adapt to stress, and report the results to outside computers. Their first use will be in critical components, such as the skins of airliners and the containment vessels of nuclear power plants, where undetected damage can be catastrophic.

Some materials will mimic the processes of life itself. Instead of being forged in great hearths or boiled in pressurized reaction vessels, they will assemble themselves from more basic substances, much as proteins assemble themselves automatically from a broth of amino acids and RNA. Here the line between material and machine, even between life and non-life, begins to blur. Self-assembling materials also will come into use in the first decade of the next century.

These and other wonders will change the world around us in the years to come, just as earlier developments in materials have changed it in the past. Yet the chances are that we will seldom notice them, any more than we give thought to the silicon from which our computer chips are made.

SUPERCONDUCTORS

In any given year, the world's generating plants churn out something over 12 trillion kilowatt-hours of electricity, enough to give an average American home its standard 100-amp service for the next 125 million years. Just under one-fourth of that is produced in the United States. In this country alone, that means operating more than 100 nuclear power plants, burning nearly 170 billion barrels of oil and 100 million tons of coal per year, and routing the nation's rivers through 3,362 hydroelectric dams. And of all that electricity, produced through vast effort, investment, and environmental sacrifice, nearly one-third simply disappears. It is not wasted, exactly. It goes to overcome the electrical resistance of long-distance transmission lines, appliance cords, and the windings of electric motors. It's as if one-third of the world's usable water vanished in overcoming the friction of the pipes it travels through.

No wonder superconductors are at the top of almost everyone's list of materials likely to change the world. In the long run, they probably will. And the transformation will begin in the next fifteen years.

Unlike normal wires, superconductors carry electricity without any resistance at all. Pour electricity into a superconducting cable, and all the power emerges at the other end. Put electricity into a loop of super-conducting wire, and it keeps circulating around the loop until you divert it to some practical use; it's a lot like having a perfectly efficient battery that can be recharged without limit. Coil a superconducting wire, and it forms the basis for (relatively) small, enormously powerful electromagnets.

Given such powers, it is easy to imagine creative uses for superconductors; visionary scientists and engineers have suggested many of them. Magnetic-levitation (maglev) trains are one obvious application. In principle, maglevs are simple. Instead of riding on wheels, the train is supported by powerful electromagnets over a roadbed that itself is paved with electromagnets. Manipulate the two magnetic fields, and the interaction drives the train forward like a linear electric motor. Because there is no contact between the train and its magnetic "rails," there is no friction; nearly all of the energy pumped into the system goes to support and propel the train. Experimental maglev trains have reached speeds of over 350 miles per hour (560 kph), using less energy than conventional railroads—which already are the most efficient ground transport available.

Superconducting magnets could make maglev trains really practical. With conventional electromagnets, too much of the power fed into the system goes to overcome the resistance of the wires. With superconductors, virtually all the electricity is put to productive use. According to the final report of the National Maglev Initiative prepared several years ago by the U.S. Department of Transportation, it would cost $40 million to $50 million per mile to build a superconducting maglev train system. Yet this technology is so energy-efficient that the effort would pay off in the busy Northeast Corridor and perhaps in other high-traffic regions.

Superconductors could work similar wonders for almost any electrical device that involves high currents or large magnets:

- Particle accelerators, magnetic mineral separators, and medical imaging systems all can be made smaller, more energy-efficient, and sometimes even cheaper by using superconducting magnets instead of conventional copper-wound magnets. The latest medical imagers already incorporate conventional superconductors; high-temperature superconductors will cut the price and operating costs of this expensive diagnostic equipment.

♦ Scientists at Osaka University and several Japanese manufacturing concerns have built the first full-size ship driven by magnetohydrodynamic (MHD) thrusters, jets of water accelerated by superconducting magnets. Though the sleek little *Yamato 1* reaches speeds of only 6 knots and is too small to carry any cargo, the designers forecast that MHD thrusters eventually will propel cargo and cruise ships at speeds of nearly 60 miles per hour (96.5 kph).

♦ An estimated 300,000 military weapons systems already use small superconducting packages in high-performance computer systems. As the price of superconductors drops, industry forecasters believe superconductor-based chips will find their way into civilian computers, producing business machines ten to one hundred times faster than anything now available.

All this could have been accomplished years ago, except for one thing: Superconductors work only when they are extraordinarily cold. The first superconductor was discovered by a Dutch physicist named Heike Kamerlingh Onnes in 1911. It was ordinary mercury, but chilled with liquid helium to only 4° above absolute zero, or 4 kelvins. (For comparison, a comfortable room temperature of 25° Celsius is 298 kelvins.) Over the next seventy-five years, investigators found other superconductors, but they also required liquid helium to reach their operating temperature. And liquid helium is horrendously expensive—too costly by far to be used in anything but such high-budget fields as scientific research, medicine, and the military.

By 1986, many scientists had resigned themselves to the idea that superconductivity can occur only in the forbidding neighborhood of absolute zero and that superconductors themselves would remain little more than a laboratory curiosity. Then a team of scientists at the IBM

Research Laboratory in Zurich discovered that a ceramic called lanthanum barium copper oxide became superconducting when chilled to the relatively benign temperature of 35 kelvins. The following year, researchers at the University of Alabama discovered another ceramic superconductor with a transition temperature—the temperature at which superconductivity begins—of only 93 kelvins. Since then, other superconducting ceramics have been discovered, with transition temperatures upward of 125 kelvins. At these temperatures, superconducting devices can be operated in a 77-kelvin bath of liquid nitrogen rather than helium. Liquid nitrogen costs only about $1 per gallon; with so-called high-temperature superconductors, many possible uses suddenly make good economic sense.

At the same time, a second critical problem has been solved. One of the strange things about superconductivity is that it disappears if you put the superconductor in a powerful magnetic field. So superconducting magnets—or electric motors or power transformers, which depend on magnetic fields—just don't work. Though engineers over the decades had figured out how to make magnets of "conventional" low-temperature superconductors, the new ceramic superconductors proved even more sensitive to magnetic fields than the niobium-tin alloys of old. However, in the last five years, through a combination of better ceramics and clever engineering, researchers have succeeded in making wires that remain superconductive in a magnetic field. This also has opened some new uses for high-temperature superconductors.

One hot market for the chilly new materials is in the long-distance power lines that make up continent-spanning electric grids. Copper cables are about 85 percent efficient; more than eight-tenths of the electricity that enters the cable arrives at its destination. But the 15 percent that is lost adds up quickly. In the United States alone, superconducting cables could save the equivalent of 25 million barrels of oil and 15 million tons of coal each year—not to mention the output of seventeen average-size nuclear reactors.

It has taken most of a decade to figure out how to make long cables out of brittle ceramic superconductors; it's a lot like trying to make flexible wire from charcoal briquets. But in the last five years or so, the job has been pretty well mastered. A Massachusetts-based company called American Superconductor now regularly spins cables up to a kilometer long from fine filaments of superconductor sheathed in metal.

Keeping the supercables cool will be surprisingly easy. High-tension wires are generally spun with a hollow channel down the center, which is filled with a dialectric oil to improve the cable's current-carrying capacity. In the new supercables, that channel will carry liquid nitrogen. The coolant costs less than one-tenth as much as the oil it will replace, and unlike the oil liquid nitrogen will cause no pollution if it escapes into the environment.

American Superconductor plans to have test cables in place by 1997. By 2010 or so, superconducting cables should be carrying electricity throughout the United States, Japan, and much of Europe.

High-temperature superconductors are already being used to store electricity as well as carry it. So-called SMES (for Superconducting Magnetic Energy Storage) devices contain a coil of superconducting wire. Charged with electricity and then switched to create a closed loop, the coils retain their electricity for as long as they remain cold—or until the power is drawn off for use. By 1997, at least two companies should have SMES devices on the market as power-backup units for computers and factory automation: American Superconductor, of Westborough, Massachusetts; and Superconductivity, Inc., of Madison, Wisconsin. By 2015, power companies will scale the devices up to store electricity during low-demand periods and switch it into the grids when power needs are high.

Electric motors use about 60 percent of the power generated in the United States. Inevitably, large motors will be another prime target for superconductive wire. According to one estimate, a 10,000-horsepower motor wound with superconducting wire would be about half as large as one made with copper wire; it would save about 1 million kilowatt-hours of electricity every year. By the turn of the century, the first 1,000-horsepower motor should have completed testing. A decade later, factories and chemical plants throughout the developed world will be converting to superconducting motors—and saving vast quantities of electric power as a result.

The machine the motors drive could run on superconducting magnetic bearings. Put a cylindrical permanent magnet over a high-temperature superconductor, and it spins easily, but is very hard to move vertically or from side to side. As a bearing, it is not quite frictionless but about 1,000 times closer to it than a good roller bearing. We will see magnetic bearings soon in flywheel energy storage systems and other devices where friction must be kept to an absolute minimum.

There is nothing dramatic or "sexy" about most of these applications, certainly nothing as attention-getting as a maglev train or MHD ship. But industrial society is built on these foundations—cables to supply electricity, bearings to reduce friction, and giant motors to power factories, irrigation pumps, oil refineries, and chemical plants. Save for the engineers among us, we will hardly notice the change as superconductors take the place of copper and aluminum in these workaday applications. Yet we will use less coal and oil, create less pollution, and in the long run live better because they have entered our world.

INTELLIGENT MATERIALS

One thing that all materials have in common is that whatever we ask them to do, they do it passively. We smelt them, forge them, mold them, give them the shape and properties that best suit our purposes—and then they just sit there. Steel and aluminum, plastic and concrete take no active part in accomplishing our ends. They do not adjust to conditions around them. They do not heal themselves when damaged or even give notice of impending failure. Thus bridges collapse without warning and the aluminum skin of airliners fatigues until sheets of it peel away in flight.

Twenty years from now, or even ten, that may no longer be true. Researchers are designing new materials that have all the properties their forebears so conspicuously lack. With built-in "nerves" and "muscles," they will sense their own condition, adapt to their surroundings, repair their injuries, and notify us when they can no longer do their jobs. These are the so-called intelligent materials. Suddenly it is becoming difficult to tell the difference between materials and machines.

Picture a suspension bridge built a couple of decades hence, its graceful supporting pillars formed of intelligent concrete. The basic structure appears to be much as you would find in a bridge from the 1970s, ordinary concrete laced with steel reinforcing bars—but look closer. Surrounding the rebars are corrosion sensors that can tell when water seeping down through the concrete has begun to eat away the crucial reinforcement. In the concrete itself, other sensors detect hidden cracks. Optical fibers carry their findings to external computers, providing a constant readout of the bridge's health. Also embedded in the concrete are stress sensors, possibly piezoelectric crystals, which produce electricity when pressed or stretched. They activate artificial "muscles," which tense to reinforce parts of the structure where the sensors find that extra strength is needed. In high winds or an earthquake, or merely under the daily pressure of heavy traffic, the fabric of the bridge adapts to the demands placed on it. And when it can no longer adapt, it warns of its deteriorating condition.

Building such a bridge still lies well beyond our ability. Yet the components that may someday make up its intelligent concrete—the stress and corrosion sensors, the artificial muscles, and certainly the data-carrying optical fibers—all are available in the laboratory today.

No fewer than four different varieties of "muscle" are available to change the shape, position, stiffness, and other characteristics of intelligent materials. The best known are the so-called shape-memory alloys, which when bent and then warmed return to their original form. The Nitinol family, made of nickel and titanium, can regain their shape after being bent and stretched up to 8 percent of their original length. In the process, they can exert powerful forces on their surroundings.

Piezoelectric materials expand or contract in response to an electric current. Unlike shape-memory alloys, they change shape almost instantly—Nitinol is far slower—but their range of motion is much smaller. A foot-long piezoelectric rod could expand or contract by little more than a tenth of an inch, compared with roughly an inch for a Nitinol rod of the same size. Piezoelectrics are relatively fragile as well; forces that shape-memory alloys could survive easily would crush piezoeletrics.

Magnetostrictive substances are a lot like piezoelectrics, but instead of responding to electric currents, they expand in a magnetic field. Their range of motion is smaller still, about 0.1 percent of their length. They are considerably more powerful than piezoelectrics.

Finally, there are some strange liquids known as electrorheological and magnetorheological fluids. These liquids contain tiny particles of magnetic or electrically sensitive material. Subject them to an electric or magnetic field, and the particles form long chains that bind the fluid together. In a few thousandths of a second, the fluid changes from a watery liquid to something with the consistency of cold molasses. It is an odd ability, but one that potentially is useful.

In fact, each of these materials has already found practical applications. Shape-memory metals are used in robot arms and micromanipulators, where smooth, even motion rather than brute strength is a priority. Piezoelectrics are found in loudspeakers and the print heads of ink-jet printers, where fast action is more important than power or range of motion. Magnetostrictive materials have found a place in motors and powerful sonar transmitters. And electro- and magnetorheological liquids form the basis for clutches, brakes, tunable vibration isolators, and similar devices.

Early experiments have already shown that these substances can be used to govern the properties of intelligent materials. For example, Nitinol wires embedded in a structure can be heated electrically, causing them to seek their original shape. As they contract, they reinforce the structure itself. In principle, they could stiffen that twenty-first-century bridge to support growing traffic loads or to compensate for any deterioration of the concrete. In another development, Craig Rogers, a pioneer in the design of intelligent materials, and his colleagues at the Virginia Polytechnic Institute and State University Center for Intelligent Materials Systems and Structures have used piezoelectrics to reinforce high-stress regions of bonded joints, lengthening their useful life by a factor of ten. As research continues, many more such successes are sure to follow.

The other necessity for intelligent materials is a kind of "nerve" to sense the structure's condition and to carry information about it to the outside world. The development of embedded sensors and data conduits is well under way.

Optical fibers are one obvious candidate to form these nerves. By monitoring the light passing through these high-purity glass fibers, engineers can tell a lot about the condition of the material in which the fibers are embedded. At their simplest, optical fibers are a perfect on/off indicator. So long as light passes through the fibers in our hypothetical bridge, all is well. But if the fibers begin to break so that light no longer shines through them, it could well mean that cracks are accumulating in the concrete and the bridge is becoming dangerously weak. More subtle indicators also are available. By measuring properties such as the intensity and polarization of the light passing through glass fibers, scientists can detect stresses, vibrations, and even magnetic fields in the fibers' environment. These techniques could be incorporated into new construction as early as tomorrow morning.

As sensors, piezoelectric plastics have some attractive features. When stretched, they give off an electrical charge that is easy to detect. Thin films of one piezoelectric, known as polyvinylidene fluoride, have already been used in pressure sensors so acute that they can distinguish between grades of sandpaper. Robot developers are working on piezoelectric "skin" that soon may make a robot's hands almost as sensitive as human hands. It is just a matter of time before engineers learn to use similar sensors to gauge the condition of large-scale construction.

Of course, there are still problems to be solved before intelligent materials become practical. One of the most pressing is how to handle all the information that could pour from, say, an intelligent building's framework. How do we continuously weigh all the moment-by-moment reports that come from millions of sensors distributed throughout a structure that could occupy a city block and stand a quarter-mile (400 meters) tall? Computer science is still struggling to cope with such challenges. Yet progress in artificial intelligence, neural nets (computers that simulate the operation of the nervous system), and other techniques for dealing with high information densities is coming rapidly. In the next ten years, computers should easily meet the demands that intelligent materials are likely to place on them.

This is one materials revolution we are likely to feel in our daily lives. Soon we may drive over highways that actively resist the formation of potholes, or at least send for construction crews when repairs are needed. We will fly in airliners whose skins alert maintenance crews when fatigue is making them unsafe. We will live in buildings whose walls automatically brace themselves against wind gusts, earthquakes, and snow loads—and warn us if they become dangerously insecure. Even simple ladders may tell us when a rung is about to break. The world will be a more convenient place when our building materials have brains of their own. The first of these materials will be available well within the next fifteen years.

SELF-ASSEMBLY

Have you seen slow-motion films of a stream of water forming itself into droplets? Each drop forms automatically, almost magically, as it breaks free of its parent stream. In a very real sense, it assembles itself. No outside help is required to "build" a drop of water.

Life works in much the same way. If you mix in water the molecules that form a cell membrane, they automatically organize themselves into the characteristic two-layered structure that surrounds the protoplasm of a cell. Drop a template of DNA into a solution of nucleic acid bases, and RNA forms spontaneously. Put the RNA into a solution of amino acids, and a protein appears. Mix a bit of RNA with the right proteins, and a whole virus will assemble itself from its component parts. This is an elegant way to build structures far more complex than any yet made by human engineers.

Scientists are working hard to copy these processes. Just over the horizon, they hope, lies a new generation of self-assembling computer chips, free from defects caused by human error; new materials with unheard-of properties; and even machines so small that they can manipulate single atoms—the subject of our next chapter.

The water drop and the cell represent two different kinds of self-assembly, one much simpler than the other. Drops of water form because of thermodynamic laws, which require structures to assume the most stable form possible to them. Pyramids do not stand on their points but on one of the broad, stable sides. Similarly, drops of water automatically take the form that places the greatest number of molecules on the inside, where they are the most stable, and the fewest at the surface, where they are less so. In accordance with physical law, small quantities of water hanging in space form themselves into spheres. (They are distorted into the familiar teardrop shape by the passing wind as they fall through the air; look at a film of water drops in experiments hanging in midair on the Space Shuttle to see perfect spheres.) Scientists refer to this as thermodynamic self-assembly.

At least one human product already forms itself by thermodynamic self-assembly: the liposome. Modeled on the cell membrane, liposomes are small, hollow spheres with a two-layered wall. The wall consists of phospholipids, a kind of fat molecule with phosphate groups at one end. The phosphate end of the molecule is hydrophilic; it attracts water. The fatty end is hydrophobic; it avoids water. When placed in solution, phospholipids automatically form two layers, with the hydrophilic phosphate groups at the outside, exposed to the water, and the hydrophobic fat groups on the inside. The layers then break up to form hollow spheres for the same reason that water forms drops: The sphere is the most stable shape available. It is that hollow that makes liposomes useful. Drugs trapped inside the sphere as the liposome forms can be delivered to the body slowly, as they leak through the lipid wall. Depending on the exact structure of the wall material, it is even possible for liposomes to smuggle a drug directly into the cell. As medicine, this is a useful technique. As engineering, it is a ringing endorsement of the principle of self-assembly.

However, thermodynamic self-assembly works only for very simple structures. Liposomes are about as complex as they get. For more complicated structures, a much more elaborate process is required. Fortunately, nature has again shown engineers the way.

The process is known as coded self-assembly. This is the method by which cells fabricate their elaborate strands of proteins and nucleic acids. In the DNA, the intricate mechanisms of the cell carry the directions for their own construction. Mix the right materials with an organism's DNA, as we just saw, and the end products appear of their own volition. At the molecular level, coded self-assembly is just as much a clockwork mechanism as thermodynamic self-assembly. Yet the coding keeps it on track as hundreds of independent molecules (in the case of proteins) or even tens of thousands of them (for DNA) find their assigned places in the final structure. Researchers are just beginning to fantasize about designing their own systems of coded self-assembly.

With all due respect, we doubt that they will succeed in the near future. But over the next decade, science will begin figuring out how to adapt nature's own manufacturing system to fit human needs. By 2010, they will be experimenting with the first primitive electronic circuits, and perhaps submicroscopic machines, that assemble themselves. Once more, they will be blurring the distinction between materials and mechanisms. And a fascinating new technology will begin to change our world.

BUCKYSTUFF

For many chemists in the narrow and often self-absorbed world of organic synthesis, it was the Holy Grail, the Golden Fleece, and Coronado's treasure rolled into one. "Think of it," they told each other between the lectures at conferences that would have put most other people to sleep. "A spherical molecule with sixty carbon atoms. What a triumph of synthetic skill! And who knows what properties it might have?" The chemist who synthesized C_{60}, as it is abbreviated, would be the toast of his colleagues, the best of the best. He would have the scientist's quiet, lab-coated equivalent of the Right Stuff!

Richard Smalley and his colleagues at Rice University got there first. They synthesized C_{60} in 1985 and named it "buckminsterfullerene," after the architect whose geodesic domes it resembles. Other scientists called it the buckyball, not only because the molecule is a sphere but because its carbon atoms are distributed on a field of pentagons and hexagons, like the pattern of a soccer ball.

Since then, researchers all over the world have begun studying fullerenes, as molecules similar to C_{60} are known. They have made buckyballs—the term has become generic these days—containing hundreds of carbon atoms. They have made one with only thirty carbons. They have made buckyballs with atoms of neon or metal trapped inside. They have made "buckytubes," cylindrical molecules one-billionth of a meter in diameter, with walls one atom thick, and nested structures of tubes within tubes. They have made twisted fullerenes reminiscent of DNA's famed "double helix." They have made flat sheets of "buckystuff," like a buckyball opened out onto a surface, but thousands of times larger. After only a decade's work, the scientific bibliography announcing these and other discoveries extends to several thousand papers. As early as 1991, the prestigious journal *Science* named C_{60} itself "Molecule of the Year."

Along the way, scientists have discovered that buckyballs are not so difficult to make after all. They occur naturally in soot; researchers manufacture them these days by vaporizing an ordinary carbon rod with a cheap arc welder. Astrophysicists have even found evidence of natural fullerenes floating in deep space.

What all those researchers have not managed to do is find practical uses for buckystuff. That seems odd, because buckyballs and their relatives have been nominated for a long list of possible applications. Fullerenes, chemists speculated, might be used to make new catalysts, plastics, superconductors, and even rocket fuel. "We are playing with the building blocks for totally new materials," Richard Smalley declared in 1991. Benzene, he pointed out, is a simple ring of six carbon atoms; yet since it was discovered in 1825, it has been transmuted into tens of thousands of drugs, plastics, dyes, and industrial chemicals.

That comparison seemed reasonable at the time, because fullerenes do have some fascinating properties. For one thing, there is that hollow in the center. It cries out to hold something. There have been many ideas about what that something should be.

A few people suggested that fullerenes might be used to contain drugs for delivery into the body, but they soon realized that the cavity is too small for anything larger than a single atom. It might enclose a radioactive isotope for cancer treatment, but thus far no one knows how fullerenes would be distributed through the body. They might carry the isotope straight to the kidney for excretion, instead of to the tumor; alternatively, they might distribute it throughout the body. Neither outcome would be useful. And, though it seems unlikely, fullerenes themselves might be toxic or might release toxic breakdown products.

In one experiment, pharmacologists at Emory University turned a fullerene itself into a drug. A buckyball, they discovered, fit exactly into the active binding site of an enzyme critical to HIV, the virus that causes AIDS. They added two charged side chains to the molecule, to link the fullerene permanently to the enzyme, and added the new drug to a culture of HIV. And the virus lost its ability to infect human cells. So far, the modified fullerene is much weaker than AZT and other drugs already available to fight AIDS. But if scientists can boost its blocking power by a factor of 1,000, this first bucky drug might yet find its way into the world's health clinics.

As organic compounds go, fullerenes are both massive and easy to ionize. So scientists at the Jet Propulsion Laboratory suggested using them as fuel for ion-drive spacecraft engines, which provide thrust by using a magnetic field to expel charged particles. Of course, that is more likely to be practical when fullerenes are made by the ton rather than in samples of a few grams.

All those carbons invite tampering. Replace some of them with other elements, say, or hang other atoms on them like decorations on a Christmas tree. The possibilities are almost endless.

Teflon, the superslick, heatproof coating of frying pans and politicians, is composed of carbon and fluorine atoms. So why not attach a fluorine to each of the many carbons in buckystuff? It should produce molecule-size ball bearings made out of something a lot like Teflon that would be the ultimate lubricant. Unfortunately, when Harold Kroto, at the University of Sussex, finally made "buckylon," the fluorine atoms reacted with water from the air to produce hydrofluoric acid, a compound corrosive enough to dissolve whatever one tried to grease. However, it seems that $C_{60}F_{60}$ might still make a good lubricant for spacecraft components, so long as they are protected from moisture until they reached the hard vacuum of orbit.

Several teams attached potassium atoms to the surface of buckyballs. At very low temperatures, the molecules became superconductors. Adding cesium and rubidium raised the transition temperature to 33 kelvins.

Thallium brought it up to 40 kelvins, the highest superconducting temperature ever seen, save among the ceramics. Yet even this is still much colder than the transition points of superconductors already available. And these particular fullerenes decay readily when exposed to the air. Studying these compounds could help scientists learn more about superconductivity in general, but at the moment it does not look like fullerenes themselves will soon make it to the marketplace.

One of the hot tickets in fullerene research these days is buckytubes. In theory, at least, buckytubes should be immensely strong. At the tiny end of the scale, they could be ideal for making the infinitesimal machines envisioned by nanotechnology, discussed in Chapter 4. Make them a few meters long, and all those lightweight, high-strength composites using graphite and Kevlar fibers would instantly be obsolete. Some nanotubes conduct electricity; they could become the wires and electronic elements for a new generation of submicrochip. Japanese scientists have made buckytubes filled with metal atoms, which thus far rank as the finest wires ever made. Mark Pederson and Jeremy Broughton, of the U.S. Naval Research Laboratory, have even designed a nanoscale electric motor made from a few minuscule buckytubes. So far, all these potential uses remain hypothetical. No one has yet made a buckytube more than a few millionths of a meter long. Scientists are still learning how to manufacture uniform batches of buckytubes; most processes yield mixed fibers of varying lengths and diameters. There is a long way to go before buckywires and buckymotors, or even buckyfilaments, become practical.

Yet there are at least a few hints that buckytubes may have practical uses, and sooner rather than later. MicroMet Technologies, of Charlotte, North Carolina, reports that its scientists have produced a new type of steel, in which the carbon atoms form buckytubes, wrapped around the iron atoms. The new metal is half again as strong as conventional steels with the same proportions of iron and carbon and three times more durable than conventional high-performance metals. It can even be produced at a cost competitive with varieties of steel already in common use.

However, the first commercial use of fullerenes is likely to be even more prosaic. Xerox has taken out several patents on the use of bucky-balls as the toner in photocopiers. Because fullerene molecules are roughly 1,000 times smaller than the particles that form ordinary toners, they should make it possible to produce copies with much greater detail than any now available. Xerox reportedly has also made colored fuller-enes for use in color copiers.

At this point, it is hard to say where fullerene research will lead. Yet in such an active field, a few years can be a long time. A decade from now, scientists will have had plenty of time to figure out how to produce pure buckyballs and buckytubes in industrial quantities, if anyone wants them. And at that point, finding uses for them will be just a matter of engineering. We would hate to guess this early in the game which bucky-materials will finally enter our homes and workplaces, or exactly what they will be doing. But it is hard to imagine a world in which fullerene products, fifteen or twenty years hence, do not play at least a limited role in our lives.

ODDMENTS

In this chapter, we have looked at just a few categories of materials that seem likely to prove important in the near future. There are literally hundreds of specific substances that could have been included—new alloys, composite materials, and plastics, each of which will incrementally change the cars, computers, and copy machines we use every day. Here are brief looks at just a few of them:

- So-called giant magnetoresistive metals—materials that change their electrical resistance dramatically in a magnetic field—soon will find their way into sensors, computer disk drives, and memory chips. They will allow us to store much more information on a disk and to make RAM chips that retain their data when the power is cut off.

- Metallocarbohedrenes—"met-cars"—are ball-like molecules that combine carbon with metals such as titanium or zirconium. Unlike buckyballs, met-cars do not get bigger when more atoms are added. Instead, they form clusters of identical, interlocking molecules. If buckyballs never pan out as chemical catalysts, met-cars just might.

- Build a buckyball out of nitrogen atoms instead of carbon? Japanese scientists think they can do it. If so, N_{60} could be stable when frozen but instantly turn to gas when exposed to heat, releasing enormous bursts of energy. It could be one of the most powerful explosives or rocket fuels available in the early twenty-first century.

- New alloys combining aluminum and lithium offer up to 30 percent more strength than ordinary aluminum and yet are just a bit lighter. NASA engineers have used one of these alloys to design a "Super Lightweight External Tank" for the Space Shuttle; it holds as much fuel as the current tank yet weighs 7,500 pounds (3400 kg) less. Next-generation airliners will probably use aluminum-lithium alloys, significantly reducing their weight and cutting down on fuel consumption.

- Every plastic we use today is based on carbon; tomorrow's may be made from sand. Materials scientist Richard M. Laine, at the University of Michigan in Ann Arbor, has discovered that heating sand with ethylene glycol—ordinary antifreeze—and an alkaline material changes the silicon into a highly reactive form. So far, Laine has used it to make high-temperature glass, fire retardants, and clear plastics that conduct electricity. This line of research has clearly just gotten started.

- "Intelligent gels" look much like any other gel, a lot like Jell-O, in fact. But they can do startling things. Some can selectively bind toxic waste, making it easy to dispose of. Others contract in electric fields, like artificial muscles. Still others can bind to a "cargo" and let go under specific conditions; they might be used, say, to protect an acid-sensitive drug from the stomach contents and release it in the more neutral environment of the small intestine. So far, scientists are still figuring out what intelligent gels can do. But MIT physicist Toyoichi Tanaka, who made the first one in the late 1970s, has already formed a company called Gel Sciences to develop and market them.

♦ It's a rare plastic that conducts electricity, but such materials do exist, and scientists all over the world are working to develop uses for them. Many products built from conductive plastics will appear in the next fifteen years. One of the most popular is polyaniline, an ordinary-looking film. It doesn't conduct electricity all that well—copper does roughly 100,000 times better—but well enough to use in the packaging that protects computer chips against static electricity or to replace the hard-to-make copper sheath around a TV cable. Scientists at the CNRS Molecular Materials Laboratory, in France, have even managed to build an all-plastic transistor. Computer chips made from them would be roughly 1,000 times slower than silicon circuits, but that is fast enough for video displays, and in large sizes they would be a lot cheaper than today's flat panels. Look for them in a computer shop near you—in ten or fifteen years.

♦ A plasma cutting torch can easily slice through a slab of steel 18 inches (4-6 cm) thick. Yet it's stopped by a thin sheet of something called Starlite. A laser array at Britain's Atomic Weapons Establishment did manage to scorch the surface slightly; but then, it was designed to simulate the heat of a nuclear blast. The temperature on the back of the tile rose by only 25° Celsius.

Invented by Maurice Ward, a hairdresser turned plastics recycler, Starlite is a cream-colored, flexible plastic with an odd, waxy feel. No one but Ward and a few family members know what it is made of. Ward says it contains twenty-one different ingredients—plastics, ceramics, and some "additives"—but he is so bent on secrecy that he refuses even to apply for a patent. Instead, he demands that would-be developers finance a joint venture—and give him 51 percent of the company. This is one deal someone is sure to take, if no one figures out its secrets first.

4

ENGINEERING,
ONE ATOM AT A TIME

Some time in the future, we will stop building automobiles, according to a few scientific visionaries. Not that we will have abandoned highway transport; as we will see in Chapter 5, we will be driving cars, or letting them drive us, for a long time to come. But we will have given up on building things, on producing our goods by hand and machine. Instead, near-microscopic robots will "grow" automobiles—or anything else— from vats of raw materials. In the early 1950s, the U.S. Atomic Energy Commission promised that nuclear power would make electricity "too cheap to meter." This new form of manufacturing could make products from computer chips to jet airplanes almost too cheap to charge for.

The same technological revolution may supplant today's medicine while bringing new powers to repair and control our own bodies. In time, science may create virus-size machines capable of tinkering with the DNA of a single human cell, while slightly larger mechanisms float through the bloodstream to chisel cholesterol from the walls of plaque-clogged arteries. Imagine being able to mix a tasteless powder—consisting of infinitesimal machines—into your orange juice to cure a cancer or ward off heart disease.

Say good-bye to global warming as well. Billions of tiny machines copied from the photosynthetic organs of plants could scavenge excess carbon dioxide from the air of a greenhouse-heated planet and "metabolize" it into free carbon and oxygen, synthetic petroleum, or even edible protein. According to some of the more daring scientists, it could happen within the next twenty years.

There is a lot more to come, these high-tech optimists predict. Take all the technological wonders you ever imagined. Add miracles that look more like magic than like science or engineering. Then make them all nonpolluting and so cheap that anyone can afford them. That is the promise of nanotechnology, the nascent study of machines only billionths of a meter—*nano*meters—across. (Think of it as ten hydrogen atoms standing side by side, or roughly how far a human hair grows in one-tenth of a second.) Soon, they tell us, the very small is going to be very big.

The Congressional Office of Technology Assessment is convinced (or was, before the Republican majority in the 1995 Congress voted to eliminate that useful but inconveniently apolitical institution). In a 1991 report on "miniaturization technologies," the agency concluded that "Those companies and nations that can successfully develop and capitalize on miniaturization developments will reap handsome rewards."

Others go even further. Nanotechnology will "bring changes as profound as the Industrial Revolution, antibiotics, and nuclear weapons all rolled up into one massive breakthrough," predicted Dr. K. Eric Drexler in a 1986 book entitled *Engines of Creation: The Coming Era of Nanotechnology* (Anchor Press/Doubleday). It was Dr. Drexler, a private researcher and sometime lecturer at Stanford University, who coined the word "nanotechnology" in the mid-1980s. Dr. Marvin Minsky, a widely revered pioneer in computer science and artificial intelligence at the Massachusetts Institute of Technology, declares that nanotechnology "could have more effect on our material existence" than the replacement of sticks and stones by metal or the harnessing of electricity.

Nanotechnology was born more than thirty years ago in the fertile mind of Caltech physicist Richard Feynman. Three days before the end of 1959, in a lecture titled "There's Plenty of Room at the Bottom," he introduced some startling ideas to the scientists gathered at the annual meeting of the American Physical Society. "What would happen if we could arrange atoms, one by one, the way we needed?" the future Nobel laureate asked. He offered several answers. We could make wires only a few atoms across, computers with millions of times more logic elements than were then available, store the *Encyclopaedia Britannica* on the head of a pin, and tailor materials with their atoms placed more flawlessly where we wanted them than in the finest crystals. He even envisioned microelectronic circuits made by evaporating alternate layers of metal and insulator onto a substrate; it's how computer chips are made today. There was no fundamental obstacle to these startling accomplishments, he pointed out. Living cells store information, move, and manufacture complex materials, all on a smaller scale than human technology has dreamed of.

Most of the brilliant physicists in Feynman's audience thought he was joking. Primitive passenger jets were still high technology then, and the first satellite was scarcely more than two years old. The state of the art in information storage was microfilm, which squeezed a page of type into an area a bit smaller than a postage stamp. Engineers could fit only a handful of electronic parts onto their primitive microchips, and school-children still learned that atoms were too small ever to be seen. What Feynman proposed was a technology so far out of reach that more than twenty years would pass before anyone—Dr. Drexler, as it happened— would take his notion seriously.

For the most part, Feynman's vision still lies beyond the limits of the possible. The tiniest transistor on the most advanced microchip yet made is still one one-hundredth the diameter of a human hair, and that is still 1,000 times larger than the components that nanotechnologists hope eventually to create.

As challenges go, that gap is as large as nanomachines will be tiny. Yet mere difficulty has not kept governments, universities, and major corporations all over the world from pouring money into the field. As early as 1989, Great Britain established a nanotechnology development program at the National Physical Laboratories and gave it a budget of nearly $20 million—a major commitment in the often pinched world of British science. Japan's Ministry of International Trade and Industry (MITI) followed with two separate nanotech projects whose ten-year budgets total some $225 million. Just what the United States devotes to nanotechnology is harder to say. Like most American research efforts, the field is so fragmented that no one knows more than a small part of what is going on—and to protect the patents they hope someday to receive, many of the major players would like to keep it that way. There are sizable research centers at Cornell University, MIT, Stanford, and the University of Michigan. IBM, AT&T, SRI International, and just about every other American company with an interest in technology has at least dipped its toe into these waters. That is an impressive turnout for a discipline where promise remains so much easier to find than substance.

MICROMECHANISMS

To date, the clearest and best-known vision of future nanotechnology belongs to Dr. Drexler. As founder of the Foresight Institute, in Palo Alto, California, Drexler labors so tirelessly to promote his dream of "molecular manufacturing" that *Science* dubbed him "the apostle of nanotechnology."

Drexler proposes to build nanomachines one atom at a time, placing each atom exactly where it belongs, just as Feynman envisioned so long ago. In fact, in Drexler's view, no matter how small a mechanism is, if it was not built atom by atom, every atom in its assigned position, it is not really nanotechnology.

Drexler's nanoworld is a clockwork place, for although his machines are far smaller than the electronic components on today's microchips, electronics appears nowhere in them. Instead, gears, wheels, bearings—all the familiar machine parts—will be reproduced in ultra-miniature. They will be built on assembly lines identical in principle to those now turning out Fords and Chevys, but on a scale so tiny that ordinary light microscopes will never see them. Even computers will be intricate assemblies of cogs and levers, the ultimate throwback to Charles Babbage's "difference engine," designed in the nineteenth century.

If Drexler succeeds, nanotechnology could give humanity almost godlike control over the physical world. If something can be imagined, Drexler believes it can be done by his perfect little machines. Instead of making the bed in the morning, he suggests, we could have submicroscopic machines recycle it—break the sheets and blankets down into their component fibers, clean each fiber one at a time, and reweave them in place to create a perfectly made bed. A single computer might contain trillions of smaller computers, putting more power on each desktop than in all the microcomputers used in this century combined. Policemen would not wear bulletproof vests; they would have them woven under their skin by robots smaller than a cell. Magic becomes technology.

But even these are baby steps on the route to a greater goal. In the end, Drexler hopes nanotechnology will unleash swarms of tiny programmable robots that he calls "replicators." Make a single prototype of any nanomachine, and replicators will assemble them by the million from a vat of raw elements, like bacteria multiplying, and at roughly the same speed and cost. Not only can we do almost anything, we can do it within hours after the designer completes the plans, for little more than the cost of the materials. Magic becomes *cheap* technology.

So far, these miracle machines exist only as conceptual drawings and computer simulations. Drexler's schematic diagram of an assembler shows a sinuous arm with a central shaft that carries chemicals to the tip for delivery to the nanomachine under construction. His "nanofilter" consists of a "sorting rotor" that turns inside the wall between two tanks. The rotor carries antibodylike binding sites for impurities. When the binding sites are exposed to the liquid to be cleansed, they seize the unwanted molecules; when they rotate into the storage tank, the impurities are kicked off by ejecting rods activated by a central cam. But all these cams and shafts remain no more than block diagrams. They are so far beyond the current state of the art that neither Drexler nor anyone else has any idea how they might operate, much less how to build them.

At the Xerox Palo Alto Research Center, just up the road from the Foresight Institute, Drexler disciple Ralph Merkle has gone a small step further: His designs for miniature bearings and gears are complete to the individual atoms that form them. His molecular bearing, scarcely 5 nanometers (nm) across, contains 2,808 atoms of carbon, hydrogen, oxygen, nitrogen, and sulfur arranged in meshing rings. A planetary gear consists of 3,557 atoms, forming a central shaft, eight transfer shafts, and an outer ring to carry their motion. More than a billion of them could hide atop the period at the end of this sentence. But like Drexler, Merkle has no idea how to turn his concepts into reality.

Under the circumstances, not everyone is as taken by Drexler's vision as his friend Merkle. Some scientists worry that gaudy forecasts of a high-tech future will leave people so disappointed with reality that the entire field of nanotechnology will fall into disrepute. Others are skeptical of Drexler himself. "The man is a flake," declares neighbor Phillip Barth, a researcher into microelectronics at Hewlett-Packard, the computer and high-tech stalwart also based in Palo Alto. But Drexler remains confident, and it seems there are fewer skeptics every day.

MANIPULATING ATOMS

One reason may be that nanotechnology is finally beginning to produce concrete results. At least, some technologies that might become its ancestors are doing so. Thus far, practical applications remain far off.

Some of the best-known successes have emerged from a device called the scanning tunneling microscope (STM). In essence, the STM is just an electrified needle—but the needle is extraordinarily fine and is maneuvered with wondrous precision. Move it along within a few atomic diameters of a surface, and the number of electrons that tunnel between them varies with the width of the gap. Plot the electrons on a screen, and you get a picture of the lumps and valleys of the surface itself. Each lump represents a single atom.

It turns out that the STM can also move atoms around, though no one is quite sure how the effect works. In 1990, physicist Don Eigler, at the IBM research laboratory in Almaden, California, used an STM to position atoms of xenon on a supercooled nickel surface. In letters like those from a dot-matrix printer, but only 5 nanometers high, he spelled out "IBM." They were dots seen round the world; the global press made Eigler's feat the most public accomplishment of nanotechnology to date.

From writing his company's initials on a piece of metal, Eigler found it a short step to the single-atom switch. The device works by shuttling a xenon atom from an STM tip to a metal surface and back. With the atom on the surface, very little current flows, and the switch is "off"; with the atom on the tip, more current flows, and the switch is "on." Single-atom switches could eventually form the ultimate in tiny computer elements, but again that is no more than theory. In practice, no one has yet suggested a use for the new switch. And, as Eigler notes, the switch itself may be tiny, but it still comes attached to a roomful of equipment. Yet it stands as one more demonstration of our growing control over the nano-world.

In theory, it could be far more. A computer memory based on STM could fit more than 10 million multivolume encyclopedias onto a single sheet of typing paper. In fact, Eigler says a 1-inch (2.5 cm) wafer might hold as much information as all the hard and floppy disks and other memory devices IBM manufactures in a year. Unfortunately, positioning an STM tip is such a drawn-out process that writing all that information would, he estimates, "take the age of the universe."

Much closer to reality is a new family of sensors based on the STM.

In their simplest form, the sensors mount an ordinary STM needle on a delicate spring over a metal surface. Any motion bends the spring and changes the flow of electrons between the tip and the surface. Researchers at the Jet Propulsion Laboratory (JPL), in Pasadena, California, have already tested STM accelerometers. They can detect an acceleration of one hundred-millionth the force of gravity—a record-breaking zero to 60 mph (97 kilometers per hour [kph]) in nine years.

Many nano-sensors will grow from STM. Already, JPL physicists have created a detector capable of sensing the heat of a light bulb at a distance of 20,000 miles (32,000 kilometers [km]). Future models may have their STM tips coated with antibodies or chemicals that combine with viruses, allergens, pollutants, or war gases. It takes at least 100,000 molecules to set off the sensors that were the state of the art before nanotechnology entered the scene. Nano-sensors eventually should detect the weight of a single large molecule adhering to an STM needle. In the future, doctors will be able to detect a single virus that causes AIDS in a patient's blood.

CONSTRUCTION BY CHEMISTRY

At least one group of scientists and engineers regularly manufactures products next to which today's nanostructures seem enormous. And they produce them by the billion, each unit exactly identical to all the rest. In *Unbounding the Future* (Morrow, 1991), Eric Drexler points out that "A mechanical engineer, looking at nanotechnology, might ask, 'How can machines be made so small?' A chemist, though, would ask, 'How can molecules be made so large?' " In fact, a growing number of chemists have been asking themselves just that question.

The answers are well worth searching for. If chemists learn to make nanomachines, they might just turn Eric Drexler's vaporous concepts into everyday reality. At a minimum, they should be able to create new materials stronger and more durable than any now available. The silicon chips at the heart of today's computers could give way to artifical materials tailored for the precise electronic properties that circuit designers need. These applications alone could spawn whole new multi-billion-dollar industries.

We encountered one of the new giant molecules in the last chapter. To nanotechnologists, buckyballs and buckytubes look a lot like molecular TinkerToys, obvious candidates to become the bricks and girders of the nanoworld. And metallo-carbohedrenes—buckyballs with a metal atom buried at the center—could provide the framework for structures yet envisioned only dimly. Richard Smalley, the Rice University chemist who discovered buckyballs in 1990, declares that the list of possible applications for these compounds "is almost without end."

Chemists are working on many other large-cluster compounds, some of which might prove even more useful than the buckyball family. At AT&T Bell Laboratories, Michael Steigerwald has created clusters containing exactly twelve atoms of selenium, a semiconductor, and twenty-three atoms of nickel. Researchers view this as a major advance, because they need to make their clusters uniform, with exactly the same size and their atoms in exactly the same locations, before they can put them to practical use, and that has proved a difficult challenge. In the long run, Stiegerwald's first small success could lead to a whole new field of semiconductor technology.

Nature has been perfecting another branch of nanochemistry ever since life began. Proteins, nucleic acids, and some other biomolecules are nearly as large as the machines that nano-engineers hope to build, and they perform many of the functions that nanotechnology requires. In photosynthesis, other biomolecules capture energy from the sun; artificial photosynthesis could someday provide the power to operate nanomachines. Some proteins offer materials ideally suited for nano-structures. Others, the enzymes, carry out chemical reactions with much greater efficiency than human technologies. Antibodies recognize and cling to specific target molecules; they seem almost custom-made for Eric

Drexler's "nanofilters." Muscle proteins contract sharply when stimulated, providing a kind of biological motor. To many scientists, the idea of inventing a whole new nanotechnology is a lot like reinventing the wheel. Why not just adapt the almost limitless supply of materials and processes that nature has already given us?

Biomolecules offer one more unique advantage over the products of human science. As we saw in the last chapter, many of them are self-assembling. Nanotechnologists would love to duplicate this behavior. Many of them are trying.

At the University of Bath, in England, biochemist Stephen Mann has been experimenting with a cagelike protein called ferritin. In the liver, ferritin traps atoms of iron and protects the body from toxic iron oxide—rust. Mann has built nanoscale magnets from ferritin. He has also used the molecule to trap iron sulfide and manganese oxide. Mann and his colleagues are now trying to tailor ferritin's chemical affinity so that they can use it to capture still other materials. Eventually, synthetic ferritins could form the basis of nanosize chemical processing factories.

Biologists at the Scripps Clinic, in La Jolla, California, have redesigned an antibody molecule to seize one atom of zinc, just as natural antibodies latch onto foreign materials in the body. Could similar molecules form part of the sorting rotors in Eric Drexler's hypothetical nanofilters? It does seem possible.

J. Fraser Stoddart has devised a complex, self-assembling structure called a rotaxane, in which a small ring-shape molecule whirls around a circular track at a rate of 18,000 revolutions per minute (rpm). Stoddart has managed to organize his rotaxanes into layered crystals and has suggested ways to control the position of the spinning bead along its track. So far, no one has suggested any possible use for any of these compounds. Yet as a proof that science can make astonishing new mechanisms at the molecular scale, it is a stunning success.

Where this work will lead, it is too soon to tell. Yet if Drexler's tiny machines ever become real, they could well be built from artificial proteins and other compounds derived from these early efforts.

Well before that, nanochemistry may spawn a new kind of electronics based on organic molecules. One result could be a new breed of computer chip with components far smaller and faster than any now available. So far, the prospects look promising.

Molecular electronics got its start in the early 1970s, with the work of Avi Aviram, a theoretical chemist working for IBM. Aviram was the first to suggest making an electronic component—a switch—from a single organic molecule. The problem was, no one had any idea how to make the molecules he needed.

By 1989, however, Aviram had put in another fifteen years of thinking, and the art of organic synthesis had come a long way. So when he proposed making an electronic switch from one kind of X-shape molecule, James Tour, of the University of South Carolina, was able to create the necessary chemical. Aviram is still working on an even more difficult problem: Molecules are so small that the two scientists thus far cannot wire them up to find out whether their switch works.

Another kind of molecular switch may get around that problem. Stanford University chemist Steven Boxer has been working with bacterial proteins that release electrons when bathed with light. In living cells, these electrons are part of the photosynthetic process. In Boxer's experiments, they are passed along, either to other proteins or to metal electrodes. A second beam of light can be used to turn the process on or off; so can an outside electrical field. The result is a kind of nano-transistor that, in theory, could replace all the components painstakingly etched onto today's microprocessor chips. They might someday form tiny, self-assembling computers smaller and vastly more powerful than anything now available.

Where will all this lead? Michael Conrad, of Detroit's Wayne State University, has one idea. A computer scientist and biophysicist, Conrad envisions a protein computer that specializes in pattern recognition, an activity in which biological brains already excel. Confronted with a pattern, the computer would release a complicated mixture of proteins, which would assemble themselves into a unique structure. To identify the pattern, sensors would monitor the activity of the enzymes used to make the structure. The enzyme mix for each pattern would be unique. Though actually making such a computer remains well beyond today's state of the art, Conrad has managed to simulate its operation on conventional computers. In theory, the protein computer works.

Japan's national nanotechnology project has an even more ambitious goal. One research group in Japan hopes eventually to build a biomolecular computer that duplicates the human brain. A few years ago, when the scheme was first announced, virtually all American scientists thought their Japanese colleagues had lost touch with reality. Today, many of them have begun to take the artificial brain seriously.

CHIP TECH

Predictably, one of the busiest research areas grows out of optical lithography, the technique developed to make microchips. Today, the process is being used to make tiny motors and gears as well as transistors—tiny, but not yet small enough to qualify as nanotechnology. A typical motor and gear train developed at the University of Wisconsin is less than one-thousandth of an inch across (about four ten-thousandths of a centimeter); another, built by engineer Mehran Mehregany, of Case Western Reserve University, is just barely visible, yet it spins at 15,000 rpm. These devices are almost inconceivably small to anyone who grew up in a world of washing machine motors. Nonetheless, they are easily 1,000 times larger than the largest of Drexler's imaginings.

Optical lithography is a lot like enlarging a photograph, but in reverse. Chip designers create a huge negative of their microcircuit and project it through a train of lenses that reduce it to chip size. Light penetrating the transparent areas of the negative strikes a thin layer of chemical on the chip surface. Where the light strikes, this "resist" hardens into a protective coating. Where the negative blocks the light, the resist can be washed away with a solvent. The electronic components are then etched into the exposed areas of the chip. Wires made by optical lithography can be as fine as 200 nm wide, one five-hundredth the width of a human hair.

Substituting electrons for light brings even smaller structures within reach. Electron lithography could theoretically create wires only 0.2 nm across, or the width of two hydrogen atoms. In practice, such fine tracings of resist break down and prevent etching below about 10 nm.

Yet even within the limits of today's chip making, engineers are already creating useful devices with mechanical parts on something edging closer to the nanoscale. Lucas Nova-Sensor, of Fremont, California, turns out the world's smallest blood-pressure sensors, half a million of them each month. Scarcely one-twentieth of an inch long and one-third as wide over all, the sensors are so small that they can be threaded through the patient's arteries all the way to the heart. Their mechanical components are etched from silicon in the same process that creates their electronics. Silicon, Lucas designer Kurt Peterson points out, is more resilient than steel and three times as strong.

Using similar methods, Texas Instruments (TI) has created the Digital MicroMirror Device, a silicon chip just over half an inch (1 cm) square. The chip's surface consists of 300,000 movable aluminum mirrors, each covering an area of less than 13 ten-millionths of a square inch (a bit less than 2 ten-millionths of a square centimeter), individually controlled by the logic circuitry beneath. Shine a light source at the mirrors, and the result is a projection television system capable of producing brighter images and greater detail than any now in use. With a screen twelve feet away, the tiny chip creates a five-foot (1.54-meter) image with VGA resolution—640 by 480 pixels, the current standard for IBM-compatible personal computers—and a contrast ratio of fifty to one. TI recently demonstrated a desktop monitor based on the chip.

At MIT, provost Mark Wrighton heads up an even more ambitious project. Wrighton and his colleagues are studying the electronic properties of individual molecules—key information if nano-engineers are ever going to work at the atomic scale. As part of the project, they need to measure the electricity generated in individual molecules in response to light, chemicals, or an external voltage. And since no one has ever built a voltmeter that can examine a single molecule, they have set out to develop one. After several years of work, it looks as if they will eventually succeed.

This is more than a research exercise. Attached to a chlorophyll molecule, the voltmeter might measure light with unprecedented sensitivity. Hooked to an antibody or enzyme, it could sense the concentration of a biologically active substance contained in a solution or even the air. Imagine a garage door that opens automatically when it senses a buildup of carbon monoxide in the enclosed space or a meter that instantly detects an infection, such as HIV. Ten years from now, both could be on sale at local stores.

QUANTUM DREAMS

Working at the nanometer level brings unusual problems. On this scale, causality disappears, and physicists grapple with probability and statistics. Matter itself becomes iffy. Particles resemble waves. Electrons "tunnel" from one side of a gap to the other without ever existing in the gap itself. Transistors no longer work. Neither do the microchips that contain them, nor most of the other electronic devices on which modern life depends.

But there is gold down where quantum theory dwells, or so nanotechnologists hope. Confine electrons to a thin enough layer of semiconductor—10 nanometers or so—and you have what is known as a quantum well, a prison that is two-dimensional, like a sheet of infinitely thin paper, as far as electrons are concerned. Electrons are free to move up or down in a quantum well and from side to side, but not forward and back. Slice a quantum well into narrow bands, and you have a quantum wire, in which electrons can move in only one dimension. Dice it into squares, and the result is a quantum dot, where electrons have no freedom at all.

This particle penology has useful implications, thanks to Werner Heisenberg's uncertainty principle. What Heisenberg said is that it is impossible to know precisely both the position and the momentum of an object—an electron, say—at the same time. The more certain one is, the less sure the other becomes; and that is not just the product of our limited senses or instruments but the nature of reality. So the closer you tie down an electron, the broader the range of momentum it may have. And the broader the range of momentum, the higher the average energy, for roughly the same reason that pockets of the same size will contain, on average, more money if they hold coins ranging from pennies to quarters than if they hold only pennies and nickels.

All that extra energy is useful. Quantum wells were invented in the early 1970s; today, they form the transistors used in satellites and efficient lasers for compact disc players. Quantum wires developed in the 1980s form experimental lasers that are even more efficient. And quantum dots?

Just what quantum dots will be good for remains to be seen, but nanotechnologists talk of wondrous new electronic and optical properties. "Crayonium," suggested by Norman Margoulus, of MIT, could absorb and emit light at whatever wavelengths the designers want, depending on the size of the quantum dots. Again, superefficient lasers come to mind. Dense arrays of dots might be wired into computers 10,000 times more powerful than today's silicon supermachines. And some scientists envision artificial "atoms" made from quantum dots, with unimaginable new properties.

Yale researcher Mark Reed, writing in *Scientific American*, speculates that

> if engineers can fabricate lattices containing millions or billions
> of quantum dots, specifying the shape and size of each one, they
> will be able to make any electronic or optical material of which

they can conceive. Emission, absorption, and lasing spectra could be precisely tailored, and a single slab of material could even be designed to contain myriad tiny computers whose interconnections and internal architecture would change to match each new problem posed to them.

Precisely tailored quantum dots are just over the horizon. Researchers will be working with them before the 1990s are out, engineers soon after. By 2005 or so, the first quantum-dot lasers could be packing data ever more densely onto optical cards and discs. After that, it seems almost anything is possible. A new brand of quantum electronics is being born. In the long run, it could rebuild our technological world in its own image, much as early radio did and microelectronics is now doing.

CONDENSED CHARGES

Another new form of electronics also grows from the quantum world; yet it deals with electrons in bulk, not one at a time. And though its products would be dozens of times *larger* than today's microcircuits, its quantum-level roots surely qualify as nanotechnology at least as well as any chip maker's designs. In the short run—say the next twenty years—it could do more to transform our world than all the rest of nanotechnology combined.

Look at the personal computer of, say, 2010. If one old engineer's vision holds up, it will be no bigger than its keyboard, with a high-resolution color screen as thick as a shirt cardboard, enough memory to store several hundred copies of the *Encyclopaedia Britannica*, and a processor roughly 1 billion times faster than today's fastest PCs. The processor "chip" itself will probably be a sheet of embossed plastic, even cheaper than silicon and easy to stamp out by the millions.

All this grows out of an innovation called condensed charge technology (CCT). Condensed charges are beadlike structures in which 100 billion or more electrons crowd together as densely as the particles in a block of metal. They were discovered and put to use by a freelance engineer named Kenneth R. Shoulders. According to Shoulders, when electrons are packed closely enough, they no longer repel each other. Instead, they attract fiercely and form clusters. No one knows why this happens. Half a dozen competing theories attempt to explain it, but none has strong evidence to support it.

Condensed charges occur naturally in sparks, but they long went unnoticed. Shoulders has found a way to generate them efficiently and put them to practical use. He believes that condensed charge devices will soon replace virtually all the conventional electrical equipment we know today, from the simplest motors to the most advanced computers. "It looks like there is nothing in electronics that you cannot do a lot better with condensed charges than with plain old vanilla electrons," he says.

If anyone else made so grand a claim, no one would listen. But this athletic, balding, sixty-something inventor has extraordinary credentials. Shoulders despises academic and corporate politics, so he seldom stays long at one institution. Over the years, he has worked at, and left, Texas A&M, Duke University, the University of Havana, MIT, and Stanford. Lacking tenure, he supports himself by, as he puts it, "making widgets." Among the widgets he made in the early 1960s were the world's first microelectronic circuits and the prototypes of the photolithography equipment now used to manufacture both microchips and primitive nanomachines.

Condensed charges are better than single electrons, Shoulders says, simply because there are so many particles in each bundle. "Everything that happens in electronics is just wiggling electrons," he says, "and the more electrons you have in motion, the faster the action will be." One of his first practical uses for condensed charges was a switch 10,000 times faster than an ordinary transistor. It could be flipped on or off in less than one-trillionth of a second, and Shoulders expected switching speed eventually to get "a lot faster than that."

Condensed charges are quick in other ways too. They move at about one-tenth the speed of light—much faster than electrons traveling through a semiconductor. Engineers are already having trouble figuring out how to get electrons from one side of today's large chips to the other without excessive delays. Design a computer to use condensed charges, and there is no reason not to make microcircuits a foot across.

Shoulders forecasts we soon will see both giant microcircuits and a wide variety of other uses for condensed charges: hundred-horsepower electric motors no bigger than the shaft it takes to carry the torque, a flat-panel computer display with all the electronics built into it, and even an X-ray machine that fits inside a hypodermic needle. "You could put it into the patient's body to irradiate a tumor without exposing the other organs to X rays," he comments. Several companies reportedly are already experimenting with relatively straightforward applications.

It should not take them long to develop production models. Instead of following wires, condensed charges stream through fine grooves in any insulator, like water through an aqueduct. Circuits for them can be made by simply pressing the necessary lines into a sheet of plastic. "Any Third World country can do it," Shoulders declares.

Shoulders and colleagues appeared several years ago before the House Committee on Science and Technology to explain condensed charge electronics. As the engineer talked, an associate fiddled with pinking shears, plastic, a bit of brass foil, and some other odds and ends. "It looked like he was cutting out paper dolls," Shoulders recalls. "But by the time we were done, he had made a working radar transmitter. And he wasn't an engineer. He was a lawyer!"

THE BOTTOM LINE

For all its promise, and all its recent successes, nanotechnology still has a long way to go. By an inconvenient law of engineering, it is a lot easier to create one-off wonders than it is to translate them into the kind of mass-market product that changes our lives.

Writing in the British journal *Nature,* Louis Brus, of Bell Laboratories, and IBM's Don Eigler point out that:

> On the nanometer scale, we simply do not have a robust, practical method for mass production. . . . In present computers, there were two outstanding inventions: the individual transistor and mass manufacturing of integrated circuits. Nanotechnology is now in the single device invention stage, and there is no clear vision of how one could practically integrate devices in a second stage.

Wired magazine recently polled five of nanotechnology's leading pioneers about the future of their embryonic science and found a wide range of opinion on the half-dozen questions they asked. Yet all five of their experts anticipated important developments in the field in the near future. All but Eric Drexler expected nanotechnology to produce its first commercial product no later than 2005, most likely either some kind of sensor or a new form of high-density computer memory; Drexler put it off until 2015. By 2009, on average—as soon as 1998 according to Syracuse University's Robert Birge, not until 2036 in the view of Donald W. Brenner, a materials scientist at North Carolina State University— government will enact laws or regulations to control the perceived risks of nanotechnology. The first practical machine capable of assembling individual atoms into a useful structure should appear by the turn of the century, according to buckyball specialist Robert Smalley; Dr. Brenner believes not until 2025; the consensus view was 2011. By 2029, we should have medical nanomachines to repair defective or damaged cells, in the panel's average estimate; however, J. Storrs Hall, a computer scientist from Rutgers University, did not expect them to appear until 2050. Yet in his own field, Dr. Hall was the group's optimist. He expected to see the first nanocomputer by 2010. In the average view, it will not arrive until 2041, and Dr. Smalley put the date as late as 2100.

Our own rule of forecasting holds that the earliest predictions usually are too hopeful, while most distant are seldom optimistic enough. So put off the first commercial products that indisputably qualify as nanotechnology until 2010 or so, but do not be surprised if the first nanocomputer appears by, say, 2025.

Twenty years from now, the real miracles—Drexler's magical self-replicating assemblers and Japan's biomolecular brain—will still lie somewhere over the horizon. Yet nanotechnology will have begun its transition from laboratory marvel to household appliance. It should be one of the most interesting developments that science ever brings us. It might even be the ultimate revolution.

5

Y O U   C A N   G E T   T H E R E

F R O M   H E R E

"America's Orient Express," the article was titled. "Within the next 10 years," it proclaimed, "we may be able to fly from New York to Tokyo in about two hours." The article appeared in the August 1986 issue of *Popular Science* under the byline of no less an authority than General Chuck Yeager, USAF (Ret.), the first man to fly an airplane faster than the speed of sound and one of the most experienced aviators in the world. General Yeager was describing the so-called *Orient Express*, the National AeroSpace Plane, whose development had just been decreed by then-president Ronald Reagan. It was to be a technological master-piece, powered by air turboramjets, arching out of the atmosphere on each flight and reaching speeds of Mach 25. A decade later, when the *Orient Express* was supposed to make its maiden flight, it has been all but forgotten.

It was not technology that did in the Orient Express. One rumor holds that it was never meant to fly at all; it was just an excuse to develop technologies for secret military or intelligence projects and was aban-doned once its real mission had been accomplished. However, the ob-vious explanation probably is the right one: Congress was not willing to pay for so expensive a project when its usefulness was open to question. The National AeroSpace Plane would have provided transportation for that tiny minority of diplomats and executives whose schedules override any thought of cost. No one else could have afforded to fly on it.

Similar issues will dominate the technology of travel for the foreseeable future. Concerns about cost, overcrowding of the roads and skies, and pollution all are beginning to reshape the world's transportation system. In the twentieth century, the automobile has dominated short- and medium-distance travel, while the airplane has captured the long-distance market. Ships have nearly vanished from passenger service. Railroads came close to vanishing altogether. In the early twenty-first century, new technologies will greatly improve cars and airplanes and perhaps ships as well. But the most revolutionary change will be the move away from cars and airplanes. Thanks in part to better technology, the world is returning to rail. This transition is already sweeping through Europe. Despite strong political and economic opposition, it is slowly gathering momentum in the United States as well. We will examine each mode of transport in turn.

BRAINPOWER BEATS HORSEPOWER

If today's engineers have their way, tomorrow's cars will be smarter, safer, and cleaner. They should be faster as well, if you look at door-to-door travel time instead of at 0-to-60 ratings at the test track. They will not let you get lost. If you need help, they will summon it automatically. Eventually, they may even steer themselves.

The secret to much of this is—we are tempted to say "of course"—information. Driven both by government pressure and by their own competition for customers, automakers are stuffing their new dream cars with electronic sensors and communications systems modeled on those of the aerospace industry. At the same time, manufacturers, government, and academic researchers all are working on ways to make the highway itself intelligent enough to help regulate the traffic that flows over it. The goal is to cram even more cars onto the world's roads while making accidents all but impossible.

If this work succeeds, it will be just in time. Drive to a job in New York each morning—or in Boston, Washington, Chicago, or Los Angeles—and you can see the future of many smaller communities. Even with car-pool-only commuter lanes to help speed the flow of traffic, bumper-to-bumper backups are a routine aggravation. Crowding is so bad that a couple of bent fenders at the wrong intersection can leave highways clogged for miles in all directions. This problem is not limited to four-across commuter arteries. On a bad day, it can take two hours just to escape from downtown Manhattan. On the weekend, when shoppers and sightseers are on the move, roads in many populous suburbs are all but impassable. At rush hour, major highways, city streets, and suburban byways alike are carrying all the vehicles they can manage, and a few percent more. In the mid-1990s, the United States already has too many cars playing bumper tag.

Highway crowding is not just annoying; it is dangerous and expensive. According to a study by the now-disbanded Congressional Office of Technology Assessment, Americans waste roughly $100 billion each year in time and gasoline while stuck in traffic. Another $75 billion to $100 billion is lost to accidents. And a study performed by researchers at the Texas Transportation Institute at Texas A&M University found that rush-hour commuters in New York, Los Angeles, and Chicago spend an extra 110 hours per person per year—almost three full work weeks—in their cars due to traffic congestion. According to one study, commuters suffer high blood pressure in proportion to the length of their daily drive.

Other countries also face automotive overpopulation. With less well-developed road systems, Mexico City, Athens, Cairo, and many other urban concentrations suffer varying degrees of highway clogging. In Singapore, the threat of traffic congestion so worried officials that they slapped a 195 percent sales tax on new cars. Now only one in ten of the island nation's citizens owns an automobile.

In the new century, this paralytic overcrowding can only get worse. The U.S. Department of Transportation estimates that the number of cars on American highways will rise almost 50 percent by 2010. There is nowhere to put that many cars. In the next couple of decades, unless something is done to prevent it, traffic will grind to a stop in and around major metropolitan areas. The United States is not quite facing gridlock from sea to shining sea, but for many Americans it will be hard to tell the difference. Other industrialized countries are only a few years behind the United States.

Then again, maybe not. Many engineers have come to believe that the worst traffic problem is not how many cars there are but how erratic they are. Drivers surge ahead and lag behind unpredictably. They dart between lanes. They ignore speed limits and road hazards. They become distracted or fall asleep. Solve those problems, and we could pack more cars into each mile of highway and move them all swiftly and safely to their destinations.

That will require major changes in the way we drive. In the next fifteen years, cars will cease to be passive tools controlled by fallible users. Instead, they will become active partners in the task of getting from one place to another without hitting anything in between. In the short run, both drivers and their cars need more information about road conditions and the behavior of other vehicles, so they can respond to them more intelligently. In a few years, efficiency may force "drivers" to give up control over their cars and let the vehicles themselves arrange their passage with each other and with the local automated highway management system.

In all the major auto building regions, government and industry have joined forces to speed this change. In the United States, the Intelligent Vehicle and Highway System project unites the Big Three auto firms— Ford, General Motors, and Chrysler—with a host of government, industry, and university research centers. European car manufacturers, electronics firms, and research institutes have already completed the Prometheus program. (The name is an acronym for Program for European Traffic with Highest Efficiency and Unprecedented Safety.) Japan's Vehicle Information Communication System focuses on providing up-to-the-moment traffic information at highway intersections. All these programs aim to produce automated traffic management systems that eventually will give highways power over the cars that travel along them.

In fact, highways have been growing more efficient since the cellular telephone appeared. Drivers who carry a cell phone can prevent a traffic jam, or at least abbreviate it, by calling for a wrecker when their car breaks down or when they see someone else obstructing the highway. Motorola has improved on this with something it calls a cellular positioning and emergency messaging unit. It uses the Global Positioning System (GPS), an array of navigation satellites lofted by the U.S. military, to identify the car's location within a few feet and transmit it along with a call for help. That way mechanics equipped to receive the locator signal can get to the scene faster.

The Global Positioning System is making highways more efficient in another way—by ensuring that drivers themselves know where they are and where they are going. Rockwell produces a GPS receiver that can guide drivers to their destinations, either giving turn-by-turn directions or displaying a map of the route on a four-inch video screen. Oldsmobile markets it as the Guidestar system, a $2,000 option. Hertz, Avis, and National Car Rental provide similar equipment in some of their rental vehicles in major cities. So far, GPS navigation systems are limited by their cost and by the difficulty of providing sufficiently detailed maps for areas larger than a city. These problems will be solved quickly. By 2000, automotive navigational systems will contain detailed maps of the country at prices under $1000. By 2005, they will be standard equipment on most new cars. New models will also notify drivers when they near a turn on their intended route. While they will do relatively little for commuters who follow the same route every day, they should keep long-distance travelers from darting across four lanes of traffic to reach an unexpected exit ramp.

Already, Lincoln-Mercury has added one more refinement to GPS-based safety. Touch the ambulance icon in the firm's RESCU (Remote Emergency Satellite Cellular Unit) system, and a voice-activated cell phone calls the Westinghouse Emergency Response Center. A built-in GPS unit automatically relays your location, so that operators at the center can dispatch a wrecker or ambulance. Soon after the turn of the century, GPS emergency beacons also will report to the local traffic management system, so that traffic can be routed around the accident.

Traffic management systems are the next step in using information to keep the highways flowing smoothly. They monitor travel conditions and attempt to guide cars to the quickest, least congested routes. A few prototypes already are in place.

One of the most sophisticated is Inform, which speeds vehicles along 130 miles of the Long Island (New York) Expressway and its connecting roads. Inform monitors police radio and broadcast traffic reports, watches the highway through thirty-four closed-circuit TV cameras, and gathers traffic information with magnetic sensors buried throughout the corridor. Warnings of congestion flash out to motorists on 101 signs programmed by radio. So do suggestions for alternative routes. At rush hour, the system even controls traffic signals at the exit ramps to keep the highway moving. According to studies, some 7 percent more cars flow through that stretch of the expressway than would be possible without Inform, and average rush-hour speeds are 13 percent higher. Over the next fifteen years, systems like Inform will be installed on high-volume routes throughout the United States and Europe.

Second-generation traffic management equips cars with an onboard navigation system that receives broadcast reports on highway conditions. One European version called the Dual-Mode Route Guidance System (DMRG) automatically chooses the fastest route to the destination for current traffic patterns. Then it tells the driver where to go, using either a moving-map display or spoken instructions. In Sweden and the United Kingdom, so-called dynamic guidance systems already warn of construction areas, congestion, and other hazards. Similar equipment is being tested in cities across the United States. Eventually these systems will contain information on public transportation routes and schedules. At that point, if all the highways are clogged, the navigation unit will suggest cutting the drive short and taking a bus or train. DMRG already can be had as an option on a few European luxury cars. By 2005, onboard navigation will arrive in the most populous urban areas of the United States. In Europe, it will be standard equipment on most new vehicles.

With "cooperative driving," transportation really starts to become automated. At this level, cars begin to recognize hazards in their surroundings. The earliest systems will warn the driver when precautions are needed to avoid an accident. Not much later, they will take action on their own.

If you have ever backed into an unseen barrier or switched lanes only to discover that another car was lurking in your blind spot, you will appreciate this new safety equipment. Blind-spot detectors scan for obstacles behind the car. TRW's prototype triggers an alarm if you signal for a lane change while its microwave radar detects a car in your way. A similar unit from Siemens uses infrared sensors and flashes warning lights built into the driver's side mirror. Equipment from other companies checks the area directly behind the car, so drivers will no longer back into a low post or crush their child's tricycle. Blind-spot detectors should be available as options on some models by the end of 1997. By 2000, cars probably will refuse to back into an obstacle or change lanes if they sense a possible conflict with other traffic.

Even sooner, radar and infrared sensors at the front of the car will operate an automatic cruise control. If you pull up too close behind another vehicle, it will back off the throttle. If the car ahead slows down, your car will slow as well. In an emergency, this primitive intelligence will even apply the brakes. Today rear-end collisions are the most common type of accident between moving vehicles. By 2010, when old clunkers without these systems are at last leaving the highway, rear-enders could be disappearing as well.

This is where true cooperation between cars, and between cars and the highway, begins to appear—and to pay off. Vehicles with automatic cruise control will space themselves as close together as is safely possible. And because antilock braking will be standard equipment by 2005 or so, that will be very close indeed. Eventually, speed controls will be tied into the traffic management system, which will set the speed limit according to driving conditions—and will prevent cars from going any faster. All this will let more cars occupy the same amount of road. It will also cut down on queuing delays and eliminate those backups where curious drivers slow down to gawk at accidents.

Many accidents occur at night and in fog, conditions that slow traffic—or at least they should—because drivers can no longer see objects far enough ahead to stop before running over them. Soon that will no longer be true. Auto equipment manufacturers are working on devices to penetrate fog and other obscuring conditions. An infrared night-vision system from General Motors already is being tested in police cars. Ford has devised a radar system that detects objects and projects their image into the driver's field of view. Cars and trucks are outlined with thick white boundaries, while smaller items such as people are displayed as diamond-shape blocks, which begin to flash as the car approaches them. Some traffic experts are concerned that these systems will be too complicated for drivers to use; pilots, after all, go through long training to master heads-up displays much like Ford's radar system and still make disastrous mistakes. Yet it seems clear that dangers hidden by night or bad weather someday will become easier to avoid. We expect practical improved-visibility systems to reach the market by 2005 or so.

At this point, the next step is obvious: Cars eventually will begin to drive themselves. It already is happening on closed tracks and in limited highway testing. At this point, Mercedes Benz probably has more successful road time with self-steering vehicles than anyone else. Its VITA II sedan used eighteen video cameras to steer through highway traffic for a total of 5000 kilometers (km) at speeds of up to 150 kilometers per hour (kph), or just over 90 miles per hour (mph). Human drivers were there for backup but never had to seize control from the machine. Despite this and other successful experiments, it is likely to be a long time before automobiles take over their own steering. In the United States, liability problems alone would prevent even the most daring manufacturer from accepting the risk of lawsuits that an accident would bring.

However, a slightly less daring piece of automation should appear on the world's highways much sooner. Add just a little more intelligence to cars, and they will be able to talk with each other as well as with the traffic management system. This enables platooning, in which several cars form themselves into a group and move down the highway as a single unit. In one recent test of platooning, four cars drove down a California highway at speeds of up to 75 mph (120 kph), separated by less than a car length. Drivers provided the steering. The cars took care of the rest. The test lasted only 18 miles (29 km), but it was an impressive beginning. A report by the General Accounting Office to the U.S. Senate Subcommittee on Transportation concluded that this kind of automation could nearly triple the capacity of existing highways. The first platoons of cars driven by the general public should be flowing down the world's highways by 2010.

FUEL FOR THE IMAGINATION

Growing intelligence is only one of the major changes now overtaking the automobile. The twentieth century was the age of the gasoline-fueled internal combustion engine; the twenty-first almost surely will be the age of . . . something else. Around the world, automobile makers, industrial associations, public interest groups, university laboratories, and government agencies are pouring effort into the search for a better way to power your next car.

Several alternatives already are on the market. Ford's Crown Victoria sedan can be had with an engine that runs on compressed natural gas (CNG) rather than gasoline. Another option lets the company's Taurus use methanol, gasoline, or a mixture of the two. Ford also has authorized several smaller firms to convert their F-series pickups and Econoline vans to CNG. Chrysler offers options that power a few otherwise standard models with CNG, methanol, or electricity. General Motors has announced that as of 1997, it will offer all of its Chevrolet S-series cars and GMC Sonoma pickups with engines capable of using two fuels—gasoline and either ethanol or methanol, depending on the version. Automakers around the world are offering similar options.

The reason to replace our gas guzzlers—and even the gas sippers—is the growing threat of air pollution. Every day, all those internal combustion engines pour out millions of tons of carbon and nitrogen oxides, particulates, and unburned hydrocarbons. According to most estimates, something over half the world's air pollution, and perhaps one-fourth of all greenhouse gases, pour out the exhaust pipes of cars and trucks. By 2000, burning fossil fuels will spew some 7 billion tons of carbon into the air every year, which makes cars and trucks a major cause of global warming.

In the immediate future, other consequences of our addiction to gasoline may be even harder to live with. In Mexico City, a stagnant bowl between mountains where 22 million people jam themselves into 522 square miles (1,360 square kilometers), air pollution is so bad that public health authorities have considered installing coin-operated public oxygen dispensers. Their colleagues in Japan already have distributed oxygen stands around the Tokyo-Yokohama area, though pollution there is not nearly so concentrated as in the fetid Mexican capital. In Europe, vehicle-exhaust limits were reduced six times between 1970 and 1996. In the United States, concerns about air quality in Los Angeles and other populous regions have led California authorities to decree that by 1998, 2 percent of the new cars sold in the state—some 18,000 vehicles—must release no pollution at all; by 2003, 10 percent of new cars must be emission-free. Failure to meet those requirements would subject automakers to fines of $5,000 for every car by which they fall short of the goal. New York and Massachusetts have enacted similar laws, and several other states have considered doing so.

The effect of cars on air quality has not gone unnoticed by engineers. For some twenty years, they have been driven to search for ways to reduce exhaust emissions without sacrificing automotive performance. Some current efforts aim to improve the fuel efficiency of relatively conventional automobiles; the less gasoline each car uses, the less pollution it generates. Many others seek to replace gasoline with cleaner-burning fuels.

VOLTS WAGONS

The Holy Grail of this quest is a practical electric car—the only vehicle that, even in theory, would produce no pollution at all. There is irony in this. In the early 1900s, 40 percent of the cars made in the United States were powered by electricity. Thomas Edison manufactured batteries for them. In Detroit alone, four fleets of electric taxis recharged at downtown stations operated by Detroit Edison. Yet electric cars died out faster than the dinosaurs because gasoline let automobiles run faster, farther, and did not require downtime to recharge.

Now electric cars are on the move again. Chrysler, Ford, General Motors, BMW, Volvo, Peugeot-Citroën, and a host of other manufacturers have built electric vehicles. Chrysler briefly marketed an electric version of its Dodge Caravan minivan. Around the world, small companies such as trans2 of Livonia, Michigan, and Solectria, of Boston, have been established to compete with the industry's lumbering giants.

In the United States alone, according to one estimate, roughly $1 billion has been spent to develop electric vehicles since the 1980s.

To date, most firms have relatively little to show for their investment. Chrysler built about sixty of its electric minivans over two years and then shut down production after selling only a single van in 1994—and that one went to an Illinois utility that wanted it for research. The company now says it will not reenter the market until at least 1998. Volkswagen makes electric-powered Golf sedans for the relatively friendly European market, but only about one per day. General Motors once hinted that it would bring its Impact two-seater to market in 1995, but the plan was scrapped when company directors learned that gearing up to manufacture it would cost $600 million, bringing little profit in return.

In performance and cost, all these vehicles have fallen far short of the standard set by their gas-engined competition. While GM's Impact sedan could accelerate to 60 mph (96 kph) in 8 seconds, most electric cars are a lot less sporty. Top speeds seldom exceed 60 mph, and ranges are limited. Ford's Ecostar electric subminivan, based on the European Escort van, is limited to about 100 miles (160 km) one way. Of late, some development projects have reported better miles-per-charge ratings. In 1995, Boston's Solectria claimed that its four-passenger Sunrise had gone 238 miles (383 km) on a single charge. Yet a recent study by the U.S. Environmental Protection Agency found that the average range of the electric cars it tested under real-world driving conditions was only 50 miles (80 km). The stop-and-go of city driving degrades the performance of electric cars just as it does that of traditional vehicles. Worse, electric power runs straight into problems that gasoline avoids. Batteries work poorly when cold, so in winter an electric car's range usually drops to half its summer maximum. And where your car heater uses energy otherwise wasted by the engine, electric vehicles must use precious power to keep their passengers warm. It takes about as much energy to heat a car in winter as to propel it, and that cuts range in half. Ford's Ecostar gets around this by including a small diesel heater, complete with its own exhaust pipe. That solves the heating problem but undermines the mini-van's claim to be nonpolluting.

The problem with all electric cars to date is the battery. Batteries just do not hold much power for their weight and cost. A gas tank is a cheap steel can, and the fuel it holds is rich with energy. A gallon of gasoline weighs about 6 pounds (just under 10 kg) and contains nearly 120,000 watt-hours of energy. A typical lead-acid battery, the kind used to start your car, that contains as much energy as a gallon of gas weighs 400 pounds (182 kilograms [kg]). Most of the alternatives offer little better performance and cost far more. Chrysler electrified its minivan by bolting 1,800 pounds (820 kg) of nickel-iron batteries into the cargo compartment. The batteries alone cost $5,000. There is one more problem as well. Where a gas tank can be refilled indefinitely, batteries survive only a limited number of charging cycles. After that, they must be replaced. To date, most batteries must be replaced after 500 to 600 charging cycles, or less than two years of daily use. Ford estimates that $6,000 batteries with a five-year life are equivalent in cost to gasoline at $3.72 per gallon. The Advanced Battery Consortium, a development group assembled by the U.S. Department of Energy, spent four years and $260 million trying to improve on these figures. Yet thus far the reality is that batteries can be cheap, powerful, or capable of many rechargings, but not all three at once. A study panel appointed by California governor Pete Wilson—admittedly not a fan of electric cars, nor of environmental strictures in general—recently concluded that only lead-acid batteries will be ready for the automotive market by the turn of the century.

All this not only limits the performance of electric cars, it threatens to price them out of their market. In the early 1990s, J. D. Powers, a leading automotive research firm, asked more than 4,000 car owners whether they might buy an electric car. Only 6 percent said they would even think about it. And those potential buyers were not willing to pay extra for environmental friendliness. They wanted to pay $2,400 less for a battery-powered car than for one addicted to gasoline. Instead, most automakers claim that electric cars will cost more than their conventional models—$10,000 to $20,000 more.

Several highly controversial reports even have questioned whether electric vehicles offer the environmental benefits claimed for them. Though the cars themselves are clean, the technologies needed to make them work are not:

Burning coal or oil to generate electricity wastes about 65 percent of the fuel's energy. Sending the power by cable to the consumer eats up 5 to 10 percent more. Overall, this is only a little more efficient than burning gasoline.

Instead of eliminating pollution, electric cars may just shift it from the car's tailpipe to the power company's exhaust stacks. Earlier studies claimed that electric cars would cut emissions by 97 percent, including the generator's effluent. Instead, recent reports estimate that they will actually cause the release of more particulates and sulfur oxides and as much as one-third more nitrogen oxides than gas-guzzlers do today. For Los Angeles, that might still be an improvement. Much of that industrial exhaust will surround power plants in the relatively unpopulated Four Corners region of Utah, Colorado, Arizona, and New Mexico, rather than in the densely packed L.A. basin. Yet it is difficult to view that as a triumph for the environmental cause.

There may be another problem as well. According to scientists at Carnegie-Mellon University, the lead and other materials released in making and recycling lead-acid batteries for electric cars may eventually cause even more environmental damage than the auto exhaust they would replace. This charge has been hotly contested by the Electric Vehicle Association of the Americas. The organization insists that the report vastly overestimated the number of batteries likely to be produced and underestimated the benefits of using recycled lead.

Yet it seems clear that switching to electric vehicles would involve some increase in certain kinds of pollution. Whether it would outweigh the pollution saved by discarding the internal combustion engine remains uncertain.

Faced with all these problems, California authorities have backed off from their formerly inflexible demand for zero-emission cars. Automakers, they say, will get partial credit for cars that merely reduce pollution. Massachusetts and New York appear to be letting their electric-car mandates die in obscurity. This is good news for the car companies, which say they would have to tack an extra $200 to $400 onto the sticker of each conventional car sold in those states to bring the price of their electrics down to reasonable levels.

All this is not to say that electric cars have reached a dead end. Solectria's Sunrise uses an advanced nickel-metal-hydride (NiMH) battery to achieve its unusually long range. Similar batteries in the General Motors Impact reportedly can be recharged to 60 percent of their capacity in just fifteen minutes. Solectria suggests that its car could be ready for production in 1997, at a showroom price of only $17,000 (in 1995 dollars). If so, it would bring electric vehicles a level of credibility they have not enjoyed since before World War I.

Of course, electricity is only one of the possible alternatives to gasoline. Many substitute fuels also promise to mitigate the evils of internal combustion. At this point, five candidates are winning most of the attention: liquefied petroleum gas (LPG), compressed natural gas, methanol, ethanol, and hydrogen. Each has one major advantage over gasoline: burning it causes less pollution. Yet each suffers from disadvantages that could easily prevent any of them from weaning much of the private-car market away from petroleum, particularly in the United States.

CNG and LPG produce only about 15 percent as much carbon dioxide as gasoline, and even less carbon monoxide and nitrogen oxides. And while they both contain less energy per pound than gasoline, CNG turns out to be cheaper than gas when bought in bulk. The downside is that most countries lack the kind of delivery network it takes to get CNG to the customer. After some ninety years in use, gasoline can be found within easy reach of almost every town in the world. CNG and LPG stations are scarce. As a result, even where these fuels are most popular, they remain uncommon. In New Zealand, probably the world's most enthusiastic user of CNG and LPG, they account for only 4 percent of the fuel sold. Japan and Italy are not far behind. In the United States, only 0.2 percent of vehicles have been converted to use CNG or LPG, and sales of the two make up only 0.4 percent of the automotive fuel market.

Methanol and ethanol avoid some of these problems, but they have drawbacks of their own. These alcohols burn cleanly, and either could be delivered by the same stations that now sell gasoline with relatively small modifications. They also are easy to make. Methanol is produced from natural gas, while ethanol is made by fermenting biomass, much like brewing beer. But neither is cheap. A mile's worth of methanol costs about 20 percent more than gasoline, while ethanol is considerably more expensive than that. Further, supplies are surprisingly limited. For the United States, the obvious source of ethanol is corn. It would take nearly half of the American corn crop each year to replace just 10 percent of the fuel that flows through the country's gas tanks.

Brazil has by far the most practical experience with ethanol. Stung by the oil shocks of the early 1970s, the country set out in 1975 to convert its automotive fleet to alcohol. Ethanol quickly became a major part of the country's fuel supply. In just ten years, fully one-third of Brazil's cars could run on ethanol alone. Yet enthusiasm for the program has cooled. It has always taken massive government subsidies to produce ethanol at reasonable prices. Even costly imported oil turns out to be more economical than cheap, home-grown ethanol. To make ethanol competitive, oil would have to be selling at about $50 per gallon.

There are environmental concerns as well. Methanol is highly toxic, and fatal doses can be absorbed through the skin. Both methanol and ethanol mix readily with water. Spilled alcohol spreads through water in minutes, killing all life wherever concentrations are high enough. Spilled oil at least floats over the water, giving emergency workers a chance to clean up the mess. Unfortunately, methanol—the cheaper and more efficient of the two fuels—is also the more dangerous. It is far from clear whether oil or methanol would be more destructive in a shipping accident. In an ocean spill, methanol might be diluted quickly enough to limit the damage. Dumped into a river, a tankerload could eradicate the ecosystem far downstream.

In many ways, hydrogen is the ideal fuel. When it burns, it yields more energy per pound than gasoline, and the only by-product is water. An old advertisement promoting its use displayed a glass of water under a slogan to the effect that "You can drink our exhaust." Hydrogen can be obtained chemically by breaking down coal, oil, gas, or biomass, such as garbage. The cleanest way to produce it is by electrolysis, simply running an electric current through water. This creates no pollution at all, save that caused in generating the power.

Unfortunately, hydrogen is expensive. Even when produced from petroleum, a cheap and dirty source, its cost would be slightly more than that of gasoline cracked from crude oil at $25 per barrel, a price oil is not likely to see for any significant period in at least the next ten years. Hydrogen from electrolysis would be three to four times as costly, and the low estimate assumes that the electricity comes from nuclear power.

Hydrogen is inconvenient as well. For storage, it must either be compressed into a tank or absorbed into a metal substrate, from which it then must be released by heating. Stored under pressure, the hydrogen needed to propel a very efficient car 250 miles (350 km) fills a 40-gallon (150 liter) tank weighing 180 pounds (82 kg), this under a pressure of about 7,000 pounds per square inch (psi). Stored as the metal hydride, hydrogen requires a container that weighs twenty-five times as much as the fuel itself. And again, of course, there is no infrastructure to produce and distribute it.

COMPROMISES WE CAN LIVE WITH

In the end, we believe cars designed to run on any of these potential power sources will be used only for special purposes, not for everyday travel by the driving public. Even today, electric cars are fine for brief trips at relatively low speeds. So that is where they will be used—in fleets of taxis, short-range delivery vans, and other such vehicles. Cars and light trucks powered by CNG, LPG, or the alcohols similarly will be limited to fleets. In fact, most electric and alternative-fuel vehicles to date have been sold for experimental fleet use. Boston Edison, Virginia Power Co., and the German postal service all are conducting or have completed fleet tests of light electric trucks. Until 2004, the U.S. National Energy Strategy of 1993 even offers companies a tax credit of $100,000 to help them install recharging stations.

One of the most interesting attempts at a light, short-range fleet vehicle is the Tulip (Transport Urban Libre Individuel Public), a buglike two-seater from Peugeot-Citroën. Powered by a 13-horsepower (hp) (9.6 kilowatts [kW]) electric motor and nickel-cadmium batteries, it has a top speed of 47 miles per hour (75 kph). It can achieve that performance because it weighs next to nothing. Its body—seats, dashboard, and all—is assembled from just five major pieces of ultralight molded composite. Tulip's range is only 35 miles (72 kph), but for Peugeot-Citroën's purposes that is enough. The company plans to station the little cars at charging centers throughout France's largest cities and rent them out for $8 per hour as a kind of self-service taxi. The average cab ride is only two or three miles. In the long run, the company would like to station somewhere between 50,000 and 100,000 of them at some 5,000 stations in Paris alone—about the number of taxis and cab stands now found in the city. No target date for Tulip's introduction has been set.

Unlike the Tulip, most private automobiles in 2010 will be hybrid designs: They will use both an internal combustion engine and one or more electric motors.

Hybrids take advantage of the fact that a well-tuned engine creates almost no pollution unless it is under a heavy load. When a car accelerates rapidly, its engine bogs down, and its exhaust fills with oxides and unburned hydrocarbons. So long as the engine turns at its optimum speed, the car runs cleanly.

Hybrids accomplish this by using the engine to drive a generator, which keeps a small battery fully charged. The battery in turn powers the electric motors to move the car. In some designs, the engine also propels the car as it cruises along the highway, while the electric motors provide extra power for acceleration. In others, it never drives the wheels directly. Either way, it always runs at its most efficient speed. This dramatically improves mileage while nearly eliminating pollution. And because the energy comes from a chemical fuel—even gasoline—hybrids offer nearly the same performance and range as conventional cars.

The Rocky Mountain Institute, in Denver, Colorado, is working to perfect a superefficient hybrid car. Hybrid-electric cars by nature are 30 to 50 percent more efficient than ordinary gasoline power, according to research director Amory Lovins. Building them from ultralight composites would cut their weight enough to make them fully ten times more efficient, he believes. If the idea works as well as computer simulations say it will, a Hypercar, as Lovins calls his concept, should get something like 150 miles to the gallon (100 kilometers per liter) and cross the United States on a single tank of gas. In the process, Lovins estimates, it would cause less pollution than the power plants needed to charge the batteries of a purely electric vehicle.

Ford already has produced a concept car, the Synergy 2010, that uses a hybrid drive train. Though not quite the Hypercar, it represents a big first step toward a practical low-emission vehicle. It features some interesting details.

The engine is a 1.2-liter, four-cylinder model with fuel injection. It puts out 45 hp (34 kW). There is no ignition system. Instead, the engine's twenty-to-one compression ratio creates enough heat to fire the cylinders spontaneously, just as in a diesel engine. Thus, it can run on gasoline or any of the obvious alternative fuels.

The car uses regenerative braking, which captures and stores the energy that conventional brakes dissipate as heat. To slow down, the driver throws a switch, and the motor on each wheel suddenly acts as a generator, absorbing the car's momentum in the process. This energy is used to spin a flywheel, which acts as a kind of auxiliary battery. To help with acceleration, the energy is drawn back out of the flywheel and fed to the motors.

Strange, angular fenders guide air smoothly around the Synergy 2010. This helps to pare the car's coefficient of drag to 0.20. If this is not the least wind resistance of any automobile yet designed, it certainly is close.

Ford engineers have kept the Synergy 2010 relatively light. It will weigh 2,913 pounds (1,324 kg) at the curb. Though nowhere near the 900 pounds (409 kg) that Amory Lovins would like to see, this is about 1,000 pounds (455 kg) lighter than today's midsize sedans.

All this should produce a very practical family car. With seating room for six and acceleration nearly as good as a conventional car, the Synergy 2010 is expected to give 80 miles (about 130 km) to a gallon of gas. And if its pollution levels are not quite as low as those of the Hypercar, they will be a big improvement over anything in the showroom today.

RIDING THE RAILS

The same concerns now beginning to transform the automobile are bringing new life to the world's once-neglected railway systems. The best cure for rush-hour traffic is not better cars but fewer cars, and (car pools being relatively rare) each person who travels by rail is someone who has left his car in the driveway. Convince long-distance travelers that trains can be fast, efficient, and comfortable, and it might even relieve those long waits to take off and land at major airports. When Amtrak cut the rail trip from New York to Washington to 2 hours 35 minutes, the demand for air shuttles plummeted. Since the Eurotunnel opened a rail link between England and France, airline capacity between London and Paris has dropped by at least 15 percent. One small commuter airline actually went out of business.

If trains do not quite banish pollution, they come close. Compared with a car that carries only the driver, trains produce only about 40 percent as much nitrogen oxide per passenger mile (or kilometer) and hardly any particulates or carbon monoxide. They use two-thirds less energy per passenger mile than a car and only one-third to one-sixth as much as an airplane, depending on the source of the estimate.

Trains are even safer. In the United States, a driver's risk of accidental death is eighteen times that of a rail passenger. In France, it is eighty times higher. Japan's famed *Shinkansen* "bullet trains" routinely go for years and billions of passenger miles without a fatal accident.

With incentives like those, countries where trains have been in decline for years have suddenly begun to rebuild their rail systems. Lands that never abandoned the train are building more. Sweden now spends as much each year to augment its railroads as it does to improve its highways. Germany spends more on rail than on roads and will continue to do so at least through 2010. Britain is adding a high-speed rail (HSR) link between London and the English Channel Tunnel. (High-speed rail systems are those that run at 150 mph—240 kph—or greater; the fastest are capable of twice that speed.) Switzerland is so worried about highway congestion that beginning in 2004, all freight trucks not planning to stop inside the country must ride through on railway flatcars. Korea is about to build its first HSR line from Seoul to Pusan. Japan has begun a $5.9 billion program to expand its bullet-train lines. The European Community has committed itself to a $76 billion program linking all of the Continent's major cities with some 19,000 miles (30,000 km) of new rail lines.

Even in the United States, where trains carry the world's smallest share of intercity travel, rail seems in for a boom. By 1999, Amtrak finally will complete its renovation of the Northeast Corridor, linking Boston and Washington with electric trains traveling at 150 mph (240 kph). Other proposed high-speed rail systems would form a dense network from Minneapolis/St. Paul in Minnesota to Topeka, Kansas, to Chicago; Washington, D.C.; Charlotte, South Carolina; and Quebec, in Canada.

Central and southern California would be tied to Reno and Las Vegas, Nevada; Phoenix and Tucson, Arizona; and Tijuana, in Mexico. Lesser systems are planned for the regions around Seattle, Washington; Denver, Colorado; Dallas/Ft. Worth, Texas; and Atlanta, Georgia. Early in 1996, Florida signed contracts for a new rail system linking Orlando and Tampa to Miami, which is due to open around 2005. Several cities are installing or expanding light-rail systems, a kind of updated trolley designed for intra-urban use. In 1993, the Clinton administration even proposed a $1.3 billion program to upgrade and expand the country's HSR lines. Congress killed the measure, in part because of political pressure from commuter airlines that feared new competition for medium-distance passengers. The proposal seems certain to reappear in one form or another until it is enacted.

Technologically, high-speed rail comes in two very different forms, each with several variations. One employs relatively conventional trains, riding on steel wheels over steel rails. Steel-on-steel HSR systems gain their speed by putting high-powered engines on lightweight, carefully streamlined trains. After years in service, they are off-the-shelf technology. The alternative is magnetic levitation, or "maglev." Maglev trains ride over, not on, a steel guide rail, lifted by powerful electromagnets. This eliminates rolling friction and makes it possible to drive the train magnetically, like a linear electric motor. In theory, the combination permits virtually unlimited speeds.

When maglev was invented some three decades ago—by James Powell and Gordon Danby, at the Brookhaven National Laboratory, and Henry Kolm, at MIT—many forecasters assumed that it would eventually take over long-distance rail travel. Today that seems unlikely, for two reasons.

Compared with conventional HSR, maglev systems are horrifically expensive. According to the U.S. General Accounting Office in 1993, it would cost between $10 million and $20 million per mile ($6.2 million to $12.4 million per kilometer) to install a conventional HSR line. Maglev's bills would add up to at least $20 million per mile and perhaps as much as $60 million. At the end of 1991, after ignoring the American-born technology for more than fifteen years, the United States authorized spending $725 million for a National Maglev Prototype Development Program. Four years later, when the program came up for review, Congress took one look at the costs and let the scheme die.

Maglev's appeal is waning for another reason as well. Its speed advantage over conventional HSR has started to disappear. On test tracks in Germany and Japan, maglev's top speed has stabilized in the neighborhood of 325 mph (525 kph), while conventional HSR has been getting faster. Wheeled trains now operate routinely at up to 185 mph (300 kph) and should reach 250 mph to 300 mph (400 kph to 500 kph) within ten years. One advanced Japanese prototype, the Super Train for the Advanced Railway of the 21st Century (STAR 21), already has reached 264 mph (425 kph), and the speed record for steel-on-steel HSR today is actually 3 mph faster than that of maglev.

Predictably, the vast majority of the world's proposed HSR systems will use wheeled carriages. Only Japan has committed itself to a maglev line, and even that is a single proof-of-concept route just 27 miles (44 km) long. In the United States, only one maglev route had been proposed for immediate construction, Florida's Tampa-to-Miami line; when the contracts were finally signed, the nod went to conventional HSR.

The fastest commercial train system today is France's venerable TGV (Train A Grande Vitesse), which is into its third generation since service began in 1981. TGV reaches 185 mph (300 kph) on regular runs and has chalked up speeds of over 300 mph (500 kph) in tests. France is expanding its TGV lines to cover 2,900 miles (4,700 km) of track. Spain adopted the TGV for its Madrid-to-Seville line, and Korea will follow suit. The contract for the new HSR route in Florida went to Bombardier, which markets TGV in North America.

Sweden's X2000 tilt trains have also proved popular. Unlike normal rail cars, which are fixed rigidly to their undercarriages, the X2000's passenger compartments bank in a turn, so that riders are not thrown to the side. In addition, the axles are steered around turns in the track rather than being fixed rigidly across the width of the car. This allows the X2000 to run faster and lets it operate on relatively unimproved track. That in turn reduces installation costs, giving it a useful advantage over the TGV when HSR is retrofitted to existing routes. Amtrak's Northeast Corridor project reportedly will use X2000 trains once the Boston-to-New York leg is open.

Germany too has its own HSR system, the InterCity Express (ICE), but it competes at something of a disadvantage. Not quite as fast as the TGV but unable to run on tracks open to the X2000, the ICE is used mostly at home. ICE trains now run between Hanover and Wurzburg and between Mannheim and Stuttgart. A few have been used as backups on Amtrak's New York-to-Washington run.

It says something about the field of transportation that most of the trains of 2010 will have been plying the world's tracks for nearly three decades. (ICE is an exception; it went into operation in 1993.) Almost by definition, transport projects are enormous. They take years to build and billions of dollars to pay for. Political opposition can be intense. Florida's new HSR line went through more than ten years of wrangling before the contracts were signed. Officials in Texas once announced that the Dallas/Ft. Worth–area HSR system would begin carrying passengers in 1996. Contracts were signed. Commuter airlines pressured local politicians. The deal fell through. Expect slow, intermittent change in the world's transportation systems, not sudden breakthroughs. In the United States especially, the miracle is not how quickly progress moves but that it moves at all.

A SEA CHANGE IN SHIPPING

Compared with other fields of transportation, shipping seems already to have made most of its changes. In recent decades, oceangoing vessels have become sleeker, more streamlined, and therefore faster and more fuel efficient. The steam engines of old have been replaced by improved diesel engines and turbines, which are more efficient still. Sextants have given way to satellite navigation. Ships have become so automated that human sailors have nearly disappeared; except on tourist cruises, where personal service is half the attraction, they could almost sail themselves. In short, the obvious technological advances have already given ships all the benefit we can expect of them. Yet there are several new, or at least relatively undeveloped, technologies that could radically alter marine transport. That transformation should begin within the next fifteen years.

Hydrofoil ships have been in use at least since the 1960s. The principle is simple: A winglike appendage extending below the hull lifts the ship's main structure over the surface of the waves, much as an ordinary wing lifts an airplane. This reduces hull drag, allowing higher speeds while burning less fuel. It also shields passengers from uncomfortable turbulence in rough seas, because the hydrofoil cuts through waves rather than rolling over them. For this reason, most hydrofoil ships to date have been either passenger carriers or high-speed naval craft, such as patrol boats. Now a consortium of Japanese ship builders is adapting them to carry cargo. Thus far, a 56-foot (17-meter) model of the vessel has been tested successfully. The full-size ships will be about 350 feet (107 meters) long.

Japanese shipbuilders also are working on air-cushion cargo ships, which float above the water's surface on a "pillow" of compressed air. These vehicles also have been in use for decades as passenger carriers, and more recently as military landing craft, but have not previously been adapted for cargo use.

Both hydrofoil and air-cushion vessels will provide better performance than conventional ships. According to current plans, they will be powered by gas turbines and driven by water jets. They should reach speeds of about 50 knots (57 mph or 92 kph), or twice the speed of wetted-hull, propeller-driven cargo ships. They will be able to carry about 1,000 tons each for distances of up to 500 nautical miles (570 statute miles or 920 km)—too short for crossing oceans, but perfect for use on trade routes within East Asia. If all goes well, the first such ships will enter service early in the next decade.

Water jets themselves are another old idea that finally seems ready for heavier duty in fast cargo and passenger vessels. Conventional ships are held back by the inefficiency of their propellers. High speeds disrupt the smooth flow of water past the propeller blades, and trying to move the ship faster does little but waste power. This is one reason marine travel creeps along at speeds in the neighborhood of 25 miles per hour (40 kph). Water jets, in contrast, work better at high speeds, but moving a large ship fast requires so much power that they have been limited to relatively small, short-range craft. However, McDonnell Douglas has produced a seagoing version of the turbine engines used in its MD-11 passenger jets, which use only 60 percent as much fuel per horsepower as smaller, less-efficient engines. Driven by these LM 6000 turbines, water jets can be used on far larger ships.

The first such freighters will ply the oceans by the turn of the century, if the plans of a firm called Fastship Atlantic, Inc., come to pass. The company hopes to build two, and perhaps four, cargo vessels that combine water jets with advanced hulls designed to remain efficient at high speeds and stable in rough water. Fully 800 feet (244 meters) long, with eight LM 6000 engines (including two for backup), the ships will cruise at speeds of 35 to 40 knots, or about 45 statute miles per hour (72 kph). That will cut shipping times across the Atlantic from eight days to less than four. If these first ships work out, we can expect to see many more of them built in the coming years.

A step beyond these second-generation water jets are the ones being tested by Mitsubishi Heavy Industries. These are magnetohydrodynamic, or MHD, thrusters. Where conventional jets use an enormous water pump to create their thrust, MHD thrusters have no moving parts. Instead, they work like a linear electric motor. Each thruster is a simple tube that contains two electrodes and is surrounded by a large electromagnet. Seawater passing through the tube becomes electrified and is expelled from the duct by the magnetic field. Readers of Tom Clancy's naval thriller, *The Hunt for Red October,* will recognize this as the supersecret "caterpillar drive" mechanism of the experimental Soviet submarine from which the novel takes its name. Unlike Clancy's fictional ship, Mitsubishi's test vessel is a 98-foot (30-meter) surface boat with a top speed of only 8 knots. But in theory MHD thrusters could propel future vessels to speeds as high as 100 knots (115 mph or 160 kph). The first such craft might conceivably carry out their shakedown voyages by 2010 or soon thereafter. If so, they will cause a revolution in shipping.

Yet the most radical development in ocean travel is not a ship at all but a sort of low-flying airplane that skims over the water at speeds of up to 450 mph (725 kph). Known as the "Ekranoplane," or "wingship," it is a Soviet invention already thirty years old. Wingships remain little known, because they were secret military equipment for most of their history. Different models were outfitted for passenger transport, antisubmarine warfare, and naval surface patrol and attack missions. The largest weigh more than 500 tons, as much as the giant American C-5A transport. They have major advantages over both ships and conventional airplanes.

Wingships "fly" at low altitude on a cushion of air trapped between the wing and the ground or ocean surface. Because they ride over the water, not in it, they can be many times faster than any ship and are unaffected by rough seas. And because they are supported on the air beneath the wing, they use less of their engines' power to generate lift from the wing and more of it to produce speed. The combination should make them the fastest, most efficient mode of long-distance travel available. A company called Aerocon, Inc., in Arlington, Virginia, is tinkering with the idea of building wingships that weigh up to 6,000 tons and are capable of traveling up to 5,000 miles (8,000 km). They could carry a cargo of some 1,500 tons, thirty times the payload of a Boeing 747. The first commercial wingships could take to the air well within the next twenty years.

6

THE LONG CLIMB BACK TO SPACE

On August 7, 1996, scientists from NASA's Johnson Space Center made a stunning announcement: Once upon a time, some 3 billion years ago, Mars very probably was home to a primitive form of life. Just conceivably, life might survive there still.

The story was not just idle speculation. In a meteorite that had originated on Mars, Dr. David McKay and members of his team had discovered microscopic structures that looked like bacteria, together with chemicals that resemble those produced by the decay of simple lifeforms. The two bits of evidence occurred so close together, within minute fractions of an inch, that it is difficult to imagine what, other than once-living organisms, might have produced them.

It was just the kind of discovery that space enthusiasts have long hoped might revitalize the moribund American space program. Predictably, it triggered a burst of interest in our neighboring planet. NASA's most ardent supporters called for a daring new round of manned missions like nothing seen since the 1960s—an expanded space station, a return to the moon, and eventually human flights to Mars itself. The space agency itself touted its own plan, already established, to send a series of unmanned spacecraft to Mars, two probes every twenty-six months (when Mars and Earth are closest together) through 2005. The first two missions had already been scheduled for launch in November and December of 1996. (A Russian probe set to depart then failed shortly after launch.) A few of the later missions may be moved up or redesigned to seek further evidence of life. Yet we suspect that in the end, the discovery that we may not be alone in a lifeless universe will do little to revive space exploration.

LIMITED GOALS

In the next fifteen years, the United States and other spacefaring nations will send a host of unmanned probes to Mars. But there will be no dramatic surprises in space. Humanity will not return to the moon. We will not establish our first Martian colony, or even commit ourselves to doing so. The near future will bring modest advances in the basic technology of space flight, but there is little reason to hope that they will reach practical use.

The United States, on its own or in collaboration with other space-faring powers, could accomplish virtually any task it set for itself, in orbit or beyond. From lofting a truly ambitious, permanently manned space station to establishing a colony on Mars, almost any conceivable goal requires no dramatic scientific breakthroughs, just engineering on the grand scale—and America does that superbly. But none of these dramatic possibilities will take place.

The government of the United States, which once sent men to the moon, long ago brought its attention back to Earth. Here, on the planet's crowded surface, it firmly remains. No one in a position to revive the flagging American space program has any interest in doing so: not Congress, which sets the federal budget; not the current occupant of the White House, who helps to mold our national priorities; not any of his possible successors. Space, it is almost universally agreed, costs too much and offers too little practical benefit. We can afford a minimum of low-budget research, if only to remind ourselves that we are a great nation that has not entirely abandoned its interest in learning. But beyond that, any ventures in space must pay their own way.

Other governments agree. Despite Russia's history of achievement in space, Europe's methodical program of unmanned launches, and Japan's occasional declarations of intent, we believe that in space, none of them will go where the United States fails to lead.

This chapter is about what is left. It probably is the shortest in this book. However, we do not mean to suggest that humanity's conquest of space has come to a halt. There is a chance that it will continue, on a new and perhaps more solid foundation. But more of that later. For now, let us continue with what is left of the space programs we have known for nearly four decades.

ROBOT PROBES

The American space program, as envisioned by the National Aeronautics and Space Administration, stands on three somewhat wobbly legs. The firmest is NASA's program of unmanned exploration, nearly two dozen relatively inexpensive space probes and Earth orbiters designed to learn as much as possible within a budget that is a shadow of its former self. The second, and much less stable, is International Space Station Alpha. The third, most forward-looking and least robust, is a series of three demonstration programs that will develop new, less costly launch technologies in the hope that once they are available, someone will find a use for them. This is not, it must be said, the kind of coherent, hard-driving program that put Tranquility Base on the moon. But given how little NASA has to work with—less than half the money it had, in constant dollars, at the peak of the Apollo program—it is a creditable effort.

At this point, the centerpiece of NASA's unmanned space program is its Mars missions. Mars Global Surveyor, launched in November 1996, will take high-resolution photographs of the planet's surface, giving our best look yet at Earth's nearest neighbor. Mars Pathfinder, which left a month later, includes a fixed lander and a mobile rover designed to study Martian surface chemistry. In 1998, they will be joined by Mars Surveyor '98 Orbiter, which will study the Martian atmosphere. At the same time, Mars Surveyor '98 Lander will study the planet's soil and surface chemistry and look for past and present-day water reservoirs. Twenty-six months later, Mars Surveyor '01 Orbiter will study Martian mineralogy and chemistry from on high. Another lander also is planned for the 2001 launch window, but its mission has not yet been set. Lander-orbiter pairs also are proposed for 2003 and 2005, but again their missions have yet to be set. NASA would like to land unmanned rovers on the planet, and perhaps even return surface samples to Earth, but what actually gets done will depend on how much money is available.

In addition to its Mars exploration, between late 1996 and autumn 2001, NASA will launch several Earth orbiters and some twenty space probes, along with nine international missions. Among the more interesting:

"Mission to Planet Earth" combines airborne remote sensing, Shuttle experiments, ground-based measurements, and at least four satellites, including the seventh member of the Landsat series; earlier Landsats have been photographing Earth's surface since 1972. The purpose of "MTPE" is to gather information on Earth's atmosphere, weather, and other geophysical data. One satellite will gather data on energy flow and pollution in the atmosphere; it should help to answer many of the remaining questions about global warming. Another will map the planet's ozone, filling in many details about another environmental concern.

The Discovery program flies with the slogan "Faster, Better, Cheaper." Several of the Mars probes originally were scheduled as Discovery missions. In the next five years, the program also will loft a series of low-budget probes to the moon and one asteroid. Lunar Prospector will depart in October 1997 to land on the moon, where it will study surface chemistry, magnetic and gravitational fields, and other basic science. And Near-Earth Asteroid Rendezvous will spend roughly one year in orbit around asteroid 433 Eros, photographing its surface and studying its physical characteristics; the probe was launched on February 17, 1996, and will arrive at its destination in February 1998.

A BELEAGUERED OUTPOST IN SPACE

These missions have survived the attention of congressional budget cutters because they are both cheap—the Discovery probes aim for budgets in the neighborhood of $150 million—and tightly focused on clear goals. The same cannot be said of International Space Station Alpha (ISSA). Formerly America's own Space Station Freedom, the manned platform has been trimmed repeatedly, its purpose constantly redefined as NASA planners have searched for a mission that could gain enough political support to ensure its survival. We are not sure they have succeeded even yet.

In its current form, ISSA is the work of thirteen countries, including Japan, Canada, Russia, and much of Europe. Its current budget is $28 billion, of which the United States will provide some $19 billion. (Canada and Japan have already spent $4 billion on the program.) The platform itself will be about twice as big as the original Freedom plan. It will have six laboratory modules rather than three, six full-time crew members rather than four, and more than 42,000 cubic feet under pressure rather than just over 23,000. Its operating plan calls for a broad range of research, including studies of new materials that can be made only in weightless conditions, fluid physics, combustion, and life sciences. Medical research should bring new insights into the workings of cardiovascular disease, osteoporosis, and balance disorders, which NASA scientists estimate affect some 90 million Americans. The first laboratory module is scheduled to reach orbit aboard the Space Shuttle in September 1998. The station should be complete some time in 2002. According to plan, it will remain in use through 2012.

That is, of course, if it is ever completed. There is no guarantee that it will be. The Superconducting Supercollider was both less expensive than the Space Station and more clearly essential to continued progress in its area of science, and after years of work the giant particle accelerator was half done. Yet Congress killed the program. The same fate could await International Space Station Alpha. For the moment, our forecast is that ISSA will fly, but we are not as confident about it as we are about most predictions.

NEXT-GENERATION VEHICLES

The final leg of NASA's program may be the shakiest, in part because it is likely to be canceled and because it makes sense only if it leads to an ambitious program of future space development. This is the RLV program, which aims to develop one or more Reusable Launch Vehicles. It involves three semiautonomous projects.

The first was the Delta Clipper, which in 1995 was upgraded to the Delta Clipper-Experimental Advanced, or DC-XA. After eight successful tests of the DC-X, the DC-XA flew for the first time on May 20, 1996. A stubby, semiconical craft about four stories high, it broke new ground on a number of fronts. Built largely by McDonnell-Douglas, the DC-X and XA were NASA's first rockets to be built largely of graphite-epoxy composite, a lightweight material that lets engineers lift more payload per pound of fuel. They also were the first to use engines with a throttle. In its maiden flight, the DC-XA rose about 800 feet (244 meters) into the air, then floated about 150 feet (45 meters) to one side, and landed vertically under perfect control. This may have been the most dramatically "different" two-minute flight in NASA's history. Four more such tests are scheduled.

The next stage in the RLV program is the "Small Booster Technology

Demonstrator," better known as the X-34. NASA began with some ambitious goals for this project. The X-34 is a two-stage vehicle with a winged, reusable booster and an expendable upper stage. Its purpose was to put modest payloads into low Earth orbit at relatively low cost and with quick turnaround between flights. Specific goals included the capacity to make at least twenty-five flights per year to altitudes of at least 250,000 feet (76,200 meters) and at speeds of at least Mach 8. The X-34 is required to land in a 20-knot crosswind and to fly at subsonic speeds through rain and fog. If controllers abort its mission, it will be able to land safely, even without power. And the average cost per flight will be only $500,000. The first flight is scheduled for the third quarter of 1998.

The RLV program's ultimate goal is the "Advanced Technology Demonstrator," or X-33. Larger than the X-34, the X-33 will be the first single-stage-to-orbit launch vehicle. If all goes well, it will require only 96 hours of refurbishing and preparation between flights. For contrast, the Space Shuttle was designed with a 160-hour turnaround in mind, a goal it has never come anywhere near meeting.

In a competition that included designs from McDonnell-Douglas/ Boeing and Rockwell, NASA chose a proposal from Lockheed Martin. Known as VentureStar, the new craft looks like a broad pyramid with stubby swept wings and a lifting body much like those of the early pre-Shuttle test gliders that flew in the 1960s and early 1970s. It includes some novel features. To save the weight of an undercarriage able to support the million-ton takeoff weight (half that of the Space Shuttle), VentureStar will take off vertically; emptied of cargo and weighing only 198,000 pounds (90,000 kg), it will land on an 8,000-foot (2400 meter) runway, shorter than those at most major airports. It will be powered by seven "aerospike" engines, an untested design in which the exhaust exits at the perimeter of a rectangular nozzle, around a central wedge, or spike. The engines, fueled by liquid hydrogen and oxygen, are controllable, so the craft will be steered by throttling back the engines on one side. There will be no crew. Flights will be governed by an autonomous computer system. The plans will not be finished until 1998, with a production model reaching full operational status around 2005. When it does go into service, Lockheed Martin promises that it will lift a 40,000-pound (18,200 kg) payload to Earth orbit at only one-tenth the Shuttle's cost.

Given all this, it should be easy to forecast that the United States soon will have a cheap, reusable launch vehicle to carry its twenty-first-century payloads into space. Unfortunately, there is less to NASA's RLV program than meets the eye. The DC-XA project ends with its final flight test. Its technology could be picked up by private industry—in fact, this was one aim of the entire RLV program—but there is no guarantee.

Worse yet, the X-34 project is in trouble. NASA has gradually changed the project's specifications so that it is no longer the simple, cheap launch vehicle once envisioned. In its current form, the X-34 would be launched from an aircraft, and it would have both a winged booster and two upper stages. As complexity grows, so do costs. Performance and target dates, meanwhile, are both slipping. NASA itself tried to kill the project in November 1995, but the White House intervened to save it. Rockwell, NASA's prime contractor for the project, backed out early in 1996, reportedly because engineers could not meet the new design requirements within NASA's $70 million budget and still allow the company an adequate profit. The program is now in the hands of Orbital Sciences, Inc., of Houston, Texas, which formerly was one of Rockwell's subcontractors and has never before handled a project of this magnitude.

The X-33 program also is vulnerable. Funding so far has been approved only for design work. Actually building the first X-33 will require both another congressional appropriation and presidential approval. The commitment to develop an operational single-stage-to-orbit launch vehicle will demand yet more decisions. And one version of X-33's authorization linked the program to successful completion of the X-34 project. Our guess is that the Reusable Launch Vehicle program will not survive long enough to build the first X-33.

The American space program faces critical deadlines in the years

2000 and 2012; what happens on those dates will decide NASA's future. In 2012, International Space Station Alpha reaches the end of its design life. It has been suggested that the Space Shuttle, which by then will exist largely to service ISSA, should be retired at the same time. If the Shuttle is to be flown much beyond 2012, it will require upgrades, including possible replacement of the solid-rocket boosters with a liquid-fueled version (at a cost of $7 billion) and addition of a fifth orbiter ($3 billion). A replacement should be cheaper to operate, but on the other hand, its buy-in cost could be at least as great as that of refurbishing the Shuttle. It is not at all clear which way the decision, which will made in 2000, will go.

For either choice to make sense, the United States must figure out what it hopes to accomplish in space, if it wishes to accomplish anything. No unmanned probes have been firmly scheduled much beyond the turn of the century. No further manned missions have been seriously considered. To date, International Space Station Alpha is the end of the line. At this point, it seems that after 2012, NASA could be restricted to launching research satellites and automated probes. Unless private industry can find a way to profit from it, manned flight may be limited to occasional maintenance missions, such as the one that repaired the Hubble space telescope. These are political rather than scientific decisions, and there is no way to know how they will turn out.

OTHER SPACEFARING NATIONS

Space programs elsewhere offer even less chance for dramatic new developments. There are a surprising number of programs scattered around the world. Australia, India, China, and even Brazil are in the space business, but only three efforts have substantial records in space: those of Russia, Japan, and the European Space Agency (ESA).

When the Soviet Union disappeared, nearly all its research budgets vanished as well. The Russian space program has been hit as hard as any.

The Mir space station survives, though cosmonauts occasionally have been forced to extend their stays because no vehicles were available to bring them back to Earth. Reconnaissance satellites still fly, though not quite as often as they once did. But research missions have been pared severely.

The most valuable remnant of the Soviet space program has turned out to be its missile factories. Ironically, in the new capitalist economy, launch vehicles developed under the Communists have become some of Russia's most marketable assets. They have proved to be simple, rugged, reliable, and considerably cheaper than the comparable Western rockets. Ten or twelve of the giant Proton boosters fly each year, many of them delivering payloads for Western corporations. Lockheed Martin has joined forces with Krunichev Enterprises and NPO Energia to market launch space on the Proton and the American Atlas boosters. Aerojet has contracted to buy NK-33 engines from Russian stocks, at about half the price of comparable Western products, and may license the design for production in the United States. Pratt & Whitney is selling the Russian RD-120 series engines.

Becoming the world's low-cost engine supplier will not make Russia a leader in space research again. Yet it could go a long way toward commercializing space. We will return to that subject later.

Even less drama can be expected from the European Space Agency. ESA has always been known for its conservative, systematic approach to space development. Nothing will change that in the near future.

Most of Europe's effort will go toward launching research probes and the occasional communications or weather satellite, just as it has always done. Its most important probes for the next few years include XMM, the X-ray Multi Mirror mission, due to fly in 1999; Rosetta, which will analyze the chemical makeup of Comet Schussman-Wachmann 3, leaving in 2003 and arriving at the comet in 2010; and FIRST, the Far Infrared and Submillimeter Telescope, to be launched in 2005. Additional launches include the latest Meteosat weather satellite and six to eight communications and data satellites, of which only three are firmly committed to launch on ESA boosters. Thus far, no missions have been scheduled beyond 2005. This program clearly remains evolutionary, not revolutionary.

After considerable debate, ESA joined work on International Space Station Alpha. Around 2002, it will loft the Columbus Orbital Facility, a general-purpose laboratory module for the space station. The following year, the European-developed Automated Transfer Vehicle—an automated cargo carrier—should enter service. And throughout ISSA's useful life, European astronauts will visit the station. These will be the Continent's only manned space ventures.

Finally, there is Ariane-5, Europe's latest launch vehicle, which is scheduled to fly before the end of 1996. It has two stages, both expendable, with the upper stage customized for each payload. It reportedly can propel up to 23 tons into low Earth orbit or 7 tons to a geostationary transfer orbit. Though designed to be cheap and highly reliable, this is a far more traditional booster than the American RLV proposals. It will remain Europe's first-line launch vehicle well beyond 2000.

That leaves Japan, which while relatively little-noticed in the West has a substantial history of space research. Since 1975, the National Space Development Agency has lofted some two dozen satellites and built four series of launch vehicles. The latest, known as the H-II, is capable of lifting a two-ton satellite into geostationary orbit, some 22,300 miles (35,900 km) out. The first Japanese astronaut, a payload specialist, carried out materials processing experiments aboard the U.S. Space Shuttle in September 1992.

In 1994, Japan's Space Activities Commission mapped out what could become an ambitious program of research and development. Though dealing more with "attitude and policy" than with specific programs, the report did sketch out some general goals to which Japan appears committed. Its Global Earth Observation System (GEOS) will be a family of satellites dedicated to meteorology and environmental research. Flights in this series are planned at least through 2020. Communications and data satellites and even satellite-based air traffic control are all on the menu. Three specific research programs aim at future launch vehicles. One will produce the J-II, the latest in the current series of conventional launchers. A second, known as HOPE, will be a winged rocket capable of taking off and landing horizontally and lifting small payloads to orbit. The other, still anonymous, could someday yield a small aerospace plane capable of carrying human passengers into orbit. None of these efforts has progressed beyond preparatory research.

This all sounds promising, and in the long run Japan's space activities could produce interesting results. However, the 1994 report makes it clear that most of these plans are little more than outlines, policy sketches meant to guide the country's response to whatever opportunities space might someday provide. Japan is simply keeping its options open. It will be happy to develop industries in space once it is clear where profits can be made. Other countries will have to take the risks of basic research.

PRIVATE ALTERNATIVES

It is beginning to look like much of that work will be carried out by private enterprise in the United States. Inspired by visions of space-based industry, and often using inexpensive Russian engines, nearly a dozen companies have been formed to provide launch services and—tentatively—to fill some other possible demands. NASA, at least, believes that opportunities are there for the taking. The agency's 1994 Commercial Space Transportation Study outlined half a dozen potential space industries that it believes show promise:

- ◆ Remote sensing—the use of satellites to provide information about crops, minerals, and other resources here on Earth—was worth $192 million in 1992; NASA believes the market would grow by a factor of six if launch costs could be reduced from $14 million per satellite to $4 million. It appears that prices could soon fall at least that far.

- Fast package delivery also requires a substantial drop in transportation prices. Yet at a launch cost of $2,000 per kilogram ($910 per pound), between 100,000 kg (220,000 lb) and 300,000 kg (660,000 lb) of cargo per year would benefit from "one-hour to anywhere" service. It now costs NASA about $12,000 to put a kilogram of cargo into space, while Ariane charges around $8,000. But a reduction of 70 or 80 percent in the next ten years may be possible.

- When costs drop to only $200 per kilogram ($91 per pound), satellite solar power could be viable for niche markets, such as the Arctic regions, where it is difficult and expensive to provide electricity by conventional means. It will be some time before this industry takes off.

- One startling industry, NASA believes, would be profitable even at today's launch costs: space burial. A company called Celestis already is offering to launch the cremated remains of the departed on board a funeral satellite. After orbiting Earth for some years, they would fall back into the atmosphere and be disposed of permanently in one last fiery display. The firm has contracted with a private launch service called Orbital Sciences to lift the satellite, but no date has been set.

- Another idea that NASA believes could survive even with current launch costs is a space theme park. In one version, it might feature virtual-reality capsules like the simulators now incorporated into high-end video games and theme park rides. But these simulators would be connected to sensors in orbit. "Passengers" would have the very realistic experience of going to space themselves, without ever leaving the safety of the ground. The one sensation missing would be true weightlessness—and given the threat of space sickness, that could be all to the good.

♦ However, the real opportunity, and the one that most of the nascent launch companies are after, is the market for communications satellites. There are some 100 such orbiters circling Earth today, and there soon could be many more. Demand is so great that NASA, Arianespace, and Russia's Proton program all have long waiting lists. The space agency's Office of Commercial Space Transportation estimates that payloads now under development will require five to ten launches per year in the 5- to 10-ton category and nine to twelve on smaller boosters. In fact, the number could be far larger. More than a dozen competing systems are poised to begin carrying the world's communications, direct television, and data traffic over the next decade, and they all plan to launch satellites between 1998 and 2005. Just two of the proposed systems—Motorola's Iridium and Teledisc, the $9 billion plan of cell-phone magnate Craig McCaw—would require a total of more than 900 satellites over five years and more than double the demand for launch capacity. And given an average useful life of only five years, those satellites would represent a recurring market for boosters. It is clear that not all such schemes will survive. Yet it seems the need for satellite launches can only grow in the years to come.

Some of the space industry giants certainly seem convinced. Sea Launch Co. LDC is a collaboration among Boeing Commercial Space, NASA's prime contractor for the space station project; Russia's RSC-Energia; NPO-Yuzhnoye in Ukraine; and Kvaerner a.s., a Norwegian shipbuilding giant. It means to offer launch services from a modified oil drilling platform to be stationed in international waters off the California coast. Operations will be guided from a 650-foot Assembly and Command Ship, now being built at Glasgow. The booster itself will be a three-stage affair that combines two of NPO-Yuzhnoye's Zenit launchers and a Block DM upper stage from RSC-Energia. These are some of the most reliable boosters now available; Block DM has scored 140 successful launches. The Sea Launch vehicle will be capable of lifting 15,000 kg (6700 lb) to low Earth orbit or 5,700 kg (2600 lb) to geostationary orbit. Hughes Space and Communications has already committed to launch ten of its satellites with the system. The first flight is scheduled for June 1998.

Kistler Aerospace will specialize in carrying smaller payloads to low Earth orbit, but using cheap, reusable boosters. The firm's K-1 launch vehicle will have two stages, both powered by inexpensive, well-proven engines now being imported from Russia by Pratt & Whitney. It will be capable of lofting a 7,000-pound (3,200 kg) load to low Earth orbit or 5,000 pound (2,300 kg) to International Space Station Alpha. The company has already ordered fifty-four engines, with delivery to begin in July 1997. Flights are scheduled to begin in mid-1998, with routine operation slated for the following year.

Kistler officials say they are "creating the 'UPS' of space." That would sound like little more than hubris, if not for the remarkable credentials of the people involved. CEO is Dr. George Mueller, former head of the Apollo program and creator of Skylab. Other officers include Aaron Cohen, who served as chief engineer for the Space Shuttle and as head of the Johnson Space Flight Center; Dale Myers, first director of the Shuttle program; and Henry Pohl, who once was chief engineer for International Space Station Alpha. Corporate directors include a retired chairman and CEO of Boeing, the cofounder of a major telecommunications firm, and a former Minority Whip from the U.S. House of Representatives. It will be interesting to watch their progress over the next few years.

Black Horse is just a design study thus far, but an exciting one. It began as the spare time project of Captain Mitchell Burnside Clapp, of the U.S. Air Force. He calls Black Horse a "one-stop-to-orbit" vehicle. It is a small winged craft vaguely reminiscent of the old X-15. With a crew of two and a payload of 1,000 pounds (454 kg), it could reach low orbit and remain there for up to a day. It would take off and land horizontally; nothing gets thrown away. The unique feature comes soon after launch. To save weight, Black Horse would leave the ground carrying all its fuel but only a few minutes worth of oxidizer. At around 40,000 feet (12.200 km), it would meet a slightly modified KC-135 tanker and refuel in midair, just as military jets do routinely. Thus replenished, it would point its nose upward and press on into space.

The system is nothing if not cheap. Clapp's target is to build a fleet of five to ten spacecraft, which could fly one Black Horse mission per day. The squadron's total operating budget would be around $100 million per year, roughly the same as that of the SR-71 Blackbird high-altitude reconnaissance aircraft. That translates to a launch cost of only $500 per pound of payload, or $1,100 per kilogram; with a larger fleet, the price could fall from there.

One way to pare the cost still further is with an imaginative two-ship mission. One vehicle carries the cargo, the other flies with only its crew. It turns out that the deadheading spacecraft arrives in space with 20,000 pounds (9,100 kg) of unused fuel. So pump the fuel into the cargo carrier and fly the companion vehicle back to Earth. The second Black Horse can then proceed to orbit with a useful load of 12,000 pounds (nearly 5,500 kg), plus its crew. If the launch cost of one vehicle carrying its standard payload comes to $500 per pound ($1,100 per kg), the cost for the combined mission drops to less than $85 per pound ($187 per kg)! And at that price, many missions become possible.

To date, Black Horse has been developed with a small grant from the Air Force. However, Clapp has just announced that he intends to leave the service to operate a company called Pioneer Rocketplane. The firm has already submitted a bid to take over NASA's X-34 project.

Another bid to cut the cost of spaceflight comes from Microcosm, a Torrance, California, firm that has designed a simple, sturdy new rocket engine. The 5,000-pound thrust Scorpius is made of composites and runs on kerosene and liquid oxygen. It is extraordinarily simple. According to the Office of Technology Assessment, the average liquid-fueled rocket engine contains some 15,000 parts. The Scorpius has only 18. The firm estimates that a small Scorpius-powered booster should cut launch costs by a factor of ten.

Late in 1995, the company successfully test-fired its engine for 200 seconds—long enough to propel at least a suborbital sounding rocket. The next major step is the Liberty Light-Lift Launch Vehicle, designed to put 2,200 pounds (1,000 kg) into low orbit at a cost of only $1.7 million. If all goes well, a medium-size booster would follow.

Thus far, there is no guarantee that either goal will be realized. Like many small start-up companies, Microcosm is strapped for cash. Yet according to corporate estimates, developing an entire family of products through the medium-lift launcher would cost only $100 million—less than the price of two medium-size launches at today's price. According to company president James Wertz, the reduction in launch expenses would save the United States alone $100 million per month.

We will close with two more efforts. Neither has nearly the credibility of Sea Launch or Kistler Aerospace nor the practical promise of Black Horse and Microcosm. Yet for sheer ambition, they are hard to top.

SpaceCub aims to build a ship that will do for rocket flight what the Piper Cub did for aviation in the 1930s—put the common man in the pilot's seat. The brainchild of Geoffrey Landis, a senior research associate at the Ohio Aerospace Institute, SpaceCub would be a do-it-yourself kit for the well-heeled amateur. Powered by yet another of those cheap, sturdy Russian engines, SpaceCub is intended to carry a single pilot on a suborbital journey much like that of the first American astronauts. It would take off and land at a normal airport, arcing up to an altitude of roughly 100 miles (160 km) and landing only twenty minutes later almost 1,000 miles (1,500 km) downrange. The joyride would not be cheap; SpaceCub's estimated cost is $500,000. Yet even at that price, it could find a ready market. The BD-10 is a supersonic jet now being built from kits by private pilots. It too costs about $500,000, and according to one report about 100 kits have been sold.

Last, there is Artemis, named for the Greek goddess of the moon, twin sister of Apollo. Conceived by Greg Bennet, a McDonnell-Douglas engineer who spends his days working with NASA on the international space station project, its goal is nothing less than to establish a permanent, private lunar colony and shuttle passengers to the moon. The hardware itself would consist of a Lunar Transfer Vehicle (LTV) for the run out to the moon; a Descent Stage and Lunar Base Core, much like the lunar lander that first put man on the moon; and an Ascent Stage, a kind of spacegoing motorcycle, that would carry astronauts back to the LTV. All this would ride into space as cargo on two Shuttle flights. Bennet would like to see the first flight leave in July 2002. At an estimated cost of $1.27 billion, the project has a big hurdle to overcome before then. Bennet's nonprofit Artemis Foundation hopes to finance the effort by selling stock, movie rights, and promotional opportunities.

We are not about to predict that Artemis will put humanity back on the moon, nor even that Black Horse or Microcosm will ever fly. But it seems clear that even the most conservative of the new ideas for private launch service will significantly cut the cost of spaceflight. That very likely would trigger a new generation of commercial research and development in orbit. If the price falls far enough, private enterprise may even find a practical use for manned spaceflight. And that could be humanity's best hope for a future in space.

What is NASA's role in all this? We simply do not know. It certainly will not still be trying to operate as a semicommercial launch service. The agency has already signed a contract with the United Space Alliance to take over Shuttle operations. The group, which includes Boeing and Lockheed Martin, will have responsibility for the craft itself, for flight and ground operations, and for all logistics. NASA will go back to research and development, the mission for which it was established and the one it still does best.

In October 1994, the space agency established the NASA Advanced Concepts Office specifically to figure out where it should go from there.

Its stated goal "is to develop and advance new, far-reaching concepts that may later be applied in advanced technology programs." The program overview goes on to state that

> We will seek uses of space with intellectual or commercial benefits that appeal to ordinary Americans as well as corporate America. We will create concepts that can take us beyond near Earth-orbit to the frontiers beyond. We will identify concepts that promise to enable truly new capabilities rather than merely incremental advances—to create new wealth rather than manage scarcity.

As always, what comes of this will depend as much on politics and money as on technology and imagination. That is seldom encouraging. We doubt that NASA will find any new purpose that can mobilize the public support needed to revitalize the American space program. But we wish them well.

7

E N E R G Y
W I T H O U T   T E A R S

Executives of the world's giant petroleum companies seldom refer to their business as the oil industry. They call it the energy industry, and with good reason. Some 30 percent of the energy used in the world enters the economy as petroleum. One-fourth comes from coal, one-fifth from natural gas. And despite the existence of large petrochemical industries, most of the oil and gas we use still is burned for energy.

Next to fossil fuels, other energy sources are relatively minor. Depending on the source of the estimate, somewhere between 5 and 13 percent of the world's energy is derived from "biomass," mostly wood fires. One-fourth of the power used in most countries is consumed as electricity; yet hydroelectric dams and nuclear reactors contribute only 6 percent each of the total energy we use. Sunlight, wind, geothermal heat, waves, and other potential sources of power all disappear into the last tenth of a percent of the world's energy resources. In all, about seven-eighths of the energy we use comes from burning something, either fossil fuels or wood.

That may not be true for much longer. In the early 1970s, two oil-price shocks and the growth of the environmental movement triggered a search for cleaner, more "sustainable" sources of energy. After twenty-five years, those efforts are beginning to pay off. A loose coalition of scientists, engineers, environmentalists, and even power company executives is building the foundations of a new energy economy. Two decades from now, the accumulating changes will have begun to affect us all.

PROBLEMS WITH PETROLEUM

Fossil fuels have reliably sustained the industrialized world for the last century. Despite the occasional warning from pessimists, we are not about to run out of them. At current rates of consumption, the world's proven reserves of oil should last for the next forty years, by which time we almost surely will have discovered still more petroleum. Natural gas will stretch this period still further. And we will not exhaust our coal supplies for several hundred years. For as long as we want them, fossil fuels will be available, and most likely affordable.

Yet alternatives are urgently needed. Our dependence on oil, coal, and even relatively clean-burning natural gas involves tradeoffs that no longer seem acceptable.

The most important consequence may be the greenhouse effect. (We will examine this subject more closely in the next chapter.) Burning any of the common fuels releases carbon dioxide (CO_2); in this respect, wood is no better than oil or coal. Once that CO_2 enters the air, it traps heat from the sun. A growing mass of evidence suggests that this carbon dioxide has begun to warm the planet and eventually may change global climates in unacceptable, perhaps disastrous ways. According to the consensus estimate, to stabilize the level of CO_2 in the air, the world will have to cut its emissions of carbon dioxide by 60 percent. To date, the only practical way to accomplish that is to reduce our dependence on fossil fuels.

It is not happening. Instead, in 2010 emissions will be 50 percent greater than they were in 1990. Most of the added CO_2 will come from economic progress in the Third World. China, already the largest user of coal, will have the world's fastest-growing economy during the early part of the next century. That expansion will be powered by the country's vast deposits of high-sulfur coal. India too is developing its coal reserves to promote its economy. Even the United States has contemplated using more coal to reduce its dependence on imported oil. In the long run, this has the makings of a global disaster.

The other problem is one of equity. To rely on fossil fuels is to accept that some people will be enormously rich, while others no less deserving will live in poverty. For a country to give its citizens a comfortable, secure life requires the development of industry and trade. These goals in turn demand energy to run factories, farm equipment, water purification equipment, and all the other machinery of modern society; they also demand capital to invest in those facilities. Our dependence on fossil fuels almost by definition guarantees that much of the world's population will never have enough of either.

The need for petroleum weighs heavily on the Third World. Three-fourths of the developing countries rely on oil to sustain what little industry they have. Virtually all of it is imported from the Middle East. With few exceptions, the countries of sub-Saharan Africa spend somewhere between 25 and 50 percent of their hard-currency earnings on imported oil. More than 85 percent of India's new debt in the 1980s, some $36.8 billion, went to buy oil. Much of Latin America is in the same position. At least twenty-nine of the world's poorest countries import more than 70 percent of their commercial energy. This burden makes it almost impossible to invest in development programs that might improve their quality of life.

Fossil fuels also require distribution. Where oil and coal are found, economic development is relatively simple. And where populations are dense, even imported energy can be delivered equitably; Japan has virtually no native energy resources, yet oil and electricity are available to all. But where the poor are scattered across vast or inaccessible tracts of land, oil trickles outward from port cities over primitive roads. Power lines are unknown. In these regions there is little hope for a comfortable, modern life. Thus in Africa, Asia, and Latin America, country life means surviving without modern energy resources and the conveniences they bring. For the rural poor, economic development will remain impossible until imported oil can be replaced by some local source of energy.

Countries euphemistically described as "less developed" are home to 80 percent of the human race. They consume little more than one-third of the world's total energy, and most of that is limited to the cities. On average, people in the Third World get by on just one-tenth as much energy per capita as those in the major industrialized nations. In Africa, Latin America, and Asia, much of the population still makes do with wood fires.

In some regions, that practice is rapidly becoming impossible. In surprising contrast to oil, which remains plentiful despite the forecast of pessimists, wood is running short in some of the areas that most depend on it. Tibet and Nepal, parts of India and Pakistan, and other heavily populated lands poor in resources have so plundered their forests that there is little left to burn. Madagascar is quickly destroying its woodlands, which are among the most biologically diverse and productive in the world. For these regions, the need for alternative sources of energy is immediate and pressing.

Fortunately, oil is not the only energy source available. Others are less polluting and infinitely "renewable"—they do not depend on burning irreplaceable fossil fuels. Some are available virtually anywhere on Earth. We spend the rest of this chapter examining these energy options.

LET THERE BE LIGHT

There are a few exceptions to the rule: Nuclear power, whether fission or fusion, extracts the binding energy of the atomic nucleus; tidal energy systems harness the moon's gravitational pull; geothermal wells tap the heat of the planetary core. But nearly all of the power we use today, or will use in the near future, is a gift from the sun. Coal and oil are what is left of photosynthetic plants after up to 600 million years of airless decay; the oldest deposits retain the energy of sunlight that arrived on Earth almost half a billion years before the dinosaurs walked among the giant ferns. Even our giant hydroelectric dams depend on solar energy to evaporate the water that falls as rain and eventually runs through our turbines. All the promising sources of "renewable" energy are simply other ways to capture the heat and light of the sun.

Photovoltaic cells probably are the neatest of them, if not the most efficient. So-called solar batteries are just a thin layer of silicon mounted on a backing of glass or ceramic. When light strikes them, it knocks electrons from their orbits around the silicon atoms and lets them flow through the material. Adding a semiconductor junction and wires to the layer of silicon is like putting a spillway at the edge of a pond; it gives the electrons somewhere to flow *to*. And flowing electrons constitute an electric current. Mount a panel of photovoltaic cells on the roof of an isolated hut in the Andes or the African outback, and within hours the owners can have electric lights, radio, television, and even running water. Around the world, an estimated 250,000 homes have been "solarized" in this fashion to date.

Photovoltaics (PV) are useful even without the baking radiance of an equatorial sun. Remarkably, Norway, bathed only in the thin subarctic light, has roughly 60,000 homes powered by solar panels. Among the contiguous region of the United States, a square meter of ground in the dimmest corner of rainy Washington state receives an average of more than 3,000 watt-hours of solar radiation each day; southern Arizona and New Mexico receive roughly twice that. This is enough to keep most households running comfortably.

Unfortunately, not all of that energy can be converted into electricity. The first silicon solar batteries, made at the Bell Laboratories in the 1950s, were woefully inefficient; just 4 percent of the light energy shined on them emerged as useful power.

Photovoltaics were expensive as well. The first PV cells to reach the market cost a whopping $600 per watt of electricity generated. At that price, they were useful only for photographic light meters and the power panels of satellites.

However, over the last twenty years, technical advances have made PV much more practical. The most efficient solar cells made to date convert nearly one-third of the incident light into electricity, and production cells have crept past 12 percent. And as efficiency has risen, prices have fallen. Until recently, PV cells have been made from expensive wafers cut from single crystals of silicon, the same highly purified material used in computer chips. More recently, scientists have learned to make them from polycrystalline silicon and even from glassy layers of amorphous silicon. Other materials have also come into use; cadmium telluride and copper indium diselenide offer high efficiency and easy fabrication. Thanks to these less costly materials and to the economies of mass production, the price of solar cells has dropped precipitously. PV cells now cost roughly $1 per watt, or roughly $0.12 per kilowatt-hour (kWh) over the useful life of the panel. At $0.10 per kWh, or slightly less, photovoltaics will be competitive with conventional generators in some of the more expensive power markets.

By 2010, we expect commercial PV panels to reach efficiencies of 15 percent or slightly more. Costs will fall to the neighborhood of $0.08 per kilowatt-hour. (One PV manufacturer, Enron Corp., has already startled the industry by offering to sell power from a 100-megawatt generator in the Nevada desert for only $0.055 per kWh, about $0.03 per kWh cheaper than the national average cost of electricity. However, the offer counted on substantial tax incentives to help the company meet that price with current technology.) At that point, even in the developed lands, power companies will find it cheaper and easier to supply residential customers with PV panels than to try to expand their large-scale generators and distribution grids.

Several developed countries already are counting on PV to meet a substantial part of the growth in their demand for electricity. Japan already pays up to two-thirds of the cost of household PV systems; according to government plans, 70,000 such systems will be installed in Japan by the turn of the century. Germany offers a 70 percent subsidy for PV installations. Italy either pays up to 80 percent of the cost or buys power from PV generators at rates up to $0.28 per kilowatt-hour. As a result, the use of solar energy is growing much more rapidly in these countries than in other industrialized nations with less favorable government policies.

Yet even fifteen or twenty years from now PV still will have its greatest impact on less developed lands, where substantial populations are spread out over large distances or where building conventional power lines is impractical for other reasons. South Africa has committed itself to bringing electricity to 2.5 million households by 2000. Virtually all of them live so far from the national electric grid that it will be cheaper to give them PV panels than to string new cables to them. More than 11,000 schools and many remote medical clinics also will receive electricity for the first time, thanks to PV generators. Kenya, Zimbabwe, India, and other Third World countries also have ambitious rural electrification programs that depend largely on PV power.

These efforts will bring relatively modern technologies to millions of people who have never had them. Electric lights are only the beginning. Over the next twenty years, many of the world's poor will receive refrigerators, both to keep food fresh and to preserve heat-sensitive medicines; radio and television to provide contact with the outside world; and satellite receivers to bring educational broadcasts from distant schools. Small, inexpensive, self-contained PV generators will power many of those advances. Even beyond the environmental benefits of clean power, we believe that bringing the world's poor and neglected rural populations into modern society will prove to be the most important contribution made by alternative-energy systems over the next two or three decades.

THE OTHER SOLAR POWER

Another way to generate renewable energy is by capturing the sun's warmth. Just run water through a black tube lying in the sun, and it gets hot enough to be useful. In the United States, solar water heaters have sprouted from roofs as far north as New Hampshire, removing that increment of power demand from the conventional electric grid. In Israel, where the sun shines 300 days per year, the law requires all buildings over six stories to use solar panels to heat their water. Nearly 1 million solar collectors have been installed there. The measure reportedly has trimmed 3 percent from the nation's electric bill.

Generating electricity from solar thermal (ST) power is not much harder. Again the collector is usually a black tube full of water or another liquid. In some systems, the hot liquid is used to operate a Stirling-cycle motor, a kind of externally heated piston engine. In others, the liquid boils at a relatively low temperature, and the hot gas is channeled through a turbine. Either way, the sun's heat is used to turn a conventional dynamo. The most efficient ST power systems can produce electricity from solar energy with conversion rates of up to 30 percent.

ST installations tend to be large, because it takes a lot of collector area to spin a dynamo efficiently. To concentrate the sun's heat, the collector itself is mounted at the focal point of one or more mirrors. For simplicity, something like 90 percent of the ST electric systems in use today mount a long tubular collector in a trough-shape mirror. More efficient installations use a giant parabolic dish or an array of mirrors focused on a collector at the center. Some use movable mirrors to gather in as much light as possible as the sun moves across the sky. Typical ST generators produce 50 to 200 megawatts of electricity. Due to their size, most ST power plants are designed to be connected to conventional distribution grids.

Already ST power can often compete in the marketplace with generators fired by oil or coal. In California, where ST generators capable of producing more than 350 megawatts of power have been installed since the mid-1980s, the cost of ST electricity is expected to fall to the neighborhood of $0.08 per kilowatt-hour by 2000. A study by the U.S. Congressional Office of Technology Assessment estimated that prices would drop to $0.06 or less per kWh shortly thereafter. If so, environmentally sensitive regions such as California and sun-baked lands such as Israel are likely to adopt ST power for much of the new generating capacity they will need in the early part of the next century.

BLOWING IN THE WIND

Roughly 40 percent of the solar energy that strikes the earth's atmosphere never makes it down to the planet's surface. Instead, it is absorbed by the air itself and helps to warm the climate to livable temperatures. This process occurs unevenly, and it is this differential heating that creates the vast movements of air that rush back and forth across the earth. All the hurricanes and tornadoes and gentle zephyrs that brush over the planet each day represent just one-fourth of 1 percent of the solar energy that penetrates to the lower atmosphere. This is one of the more obvious potential sources of renewable energy.

At first, it seemed a poor one. Early wind generators were costly, inefficient, and prone to break down. As a result, wind energy cost more than $1 per kilowatt-hour as recently as 1981. Since then, however, new materials have made wind turbine blades stronger and lighter, while more sophisticated aerodynamic designs have made them more efficient. Power electronics have become cheaper as well. And engineers have learned a lot about where to put wind generators for the best energy efficiency. As a result, where the winds are strongest, power is now available for as little as $0.043 per kWh. According to one estimate, the price will drop another cent or more by the turn of the century. That makes wind energy as cheap as hydroelectric dams, which are the least expensive conventional sources of electricity.

Predictably, wind has become one of the fastest-growing areas of alternative energy. In the United States, California has installed more than 1,700 megawatts of wind capacity, roughly 1.5 percent of the state's electric power and enough to supply the city of San Francisco. Major wind development projects have been announced in Washington state, Texas, and several other areas. As a whole, the United States has more than half of the world's installed capacity of wind power. Minnesota and Wisconsin, Maine and Vermont, New York, Oregon, Wyoming, and Montana all have contemplated installing wind generators. In theory, wind power from just three states—Texas and North and South Dakota—could supply the country's entire electric demand, without intruding on environmentally sensitive areas.

Other countries seem even more eager to harness the wind. Denmark began installing wind turbines in the 1970s and by 1994 had more than 3,600 in place. (Danish companies also supplied about half of the 15,000 turbines installed in California.) Germany and Great Britain each have announced plans to install wind generators capable of producing at least 1,000 megawatts of electricity by 2005; Germany already is halfway there. Members of the European Union will install 4,000 megawatts of wind generating capacity by the turn of the century and plan to double that total only five years later. Argentina, China, Ireland, New Zealand, Switzerland, and Ukraine all have announced major wind development plans. Collectively, these projects will more than triple the amount of electricity harvested from the world's winds in the next ten years.

Technologically, the wind turbines of 2010 are likely to be very similar to those of 1996. Wind power is not yet mature, but it seems that most of the field's major problems have already been solved.

There are two kinds of wind turbine. By far the most popular use blades like a giant airplane propeller, rotating on a horizontal axis. They are efficient, easily scaled up, and familiar. Their one disadvantage is that they must face into the wind. Where wind direction varies erratically, vertical turbines are used. These use an array of long, slender airfoils attached at top and bottom to a vertical axis, like the blades of an egg-beater. They collect wind coming from any direction but are not quite so easily enlarged. Some 90 percent of the world's existing turbines are of the horizontal variety. Vertical turbines will be more popular in 2010 than they are today, because generator farms will be expanding to less-than-ideal locations. Yet horizontal turbines will continue to dominate the field.

The important changes will be in the details: airfoils used to collect the wind's energy; stronger, lighter materials; and more versatile power electronics. One problem with most of today's turbines is that they re-quire steady winds. Outside a relatively narrow range of speeds, their airfoils are no longer efficient. If high winds turn them too quickly, they tend to break. And their frequency depends on their propeller speed. If you want 50 Hertz (Hz) (in Europe) or 60 Hz (in the United States) AC power, the turbine must turn at the proper rate. Solutions to all these problems are available today. Engineers have developed new airfoils that let turbines extract more of the wind's energy over a wider range of speeds. These designs also limit the turbine's rotational speed, so there is less risk of damage. And more sophisticated power electronics make it possible to generate the proper current regardless of turbine speed. As today's wind farms age, these innovations will be retrofitted to exist-ing wind turbines. New wind farms will incorporate them during con-struction. By 2010, most of the turbines in use will benefit from these technologies.

There is one more way to deal with the inconvenient variability of natural breezes: Make your own wind. The idea was first proposed in 1975 by an American scientist named Philip Carlson. It is being devel-oped by researchers at Israel's Technion Institute of Technology, in Haifa. The idea is to build a huge, hollow tower in a convenient desert; hot and dry surroundings are a must. Then pump sea water to the top of the tower and spray it into the center of the tube. Evaporation will cool the warm, dry air at the top of tower and cause it to plummet toward the ground. Just mount a turbine at the bottom of the tower, and use this artificial wind to spin your generators.

These energy towers, as the Technion group calls them, theoretically could solve the world's energy problems. According to calculations, the planet offers enough hot, dry air to produce many times more energy than the world now uses, at a cost as low as $0.007 per kilowatt-hour, less than one-tenth the average price of electricity in the United States.

The only obstacle is their size. A practical energy tower must be close to 3,000 feet (900 meters) high and would cost roughly $650 million to build. Even a small demonstration project is still in the talking stage, and scaling up such a prototype could involve so many unforeseen engineering problems that it is likely to be a long time before anyone commits to this concept.

Wind power will be cheaper and more practical ten or fifteen years from now, not because the world has built energy towers but because of the more humble improvements now being made in conventional wind turbines. Many of the early wind farms existed only because tax incentives made them possible. When those incentives were repealed, their developers went bankrupt. This was particularly common in California, where many wind turbines have survived their original owners. By 2010, where winds are favorable and the cost of conventional energy is high, wind power will be able to survive in direct competition with fossil fuels.

COPING WITH BROWN-OUTS

Solar energy, both photovoltaic and thermal, has some drawbacks; so does wind energy. Solar power is not very practical at night or when skies are overcast. Winds vary with the season and time of day; their peak availability may not coincide with the maximum demand for electricity. And often we need to store energy—in our gas tanks, for example—for use away from the generator and the electric power grid.

Solar and wind energy do not lend themselves to storage to meet remote demands. To overcome these limitations, engineers have devised a variety of solutions.

Small photovoltaic systems of the kind now bringing electricity to remote corners of the Third World include batteries to store power for use at night or on dim days. The systems work well enough in such free-standing installations, yet batteries are expensive. In small PV systems, they dramatically raise the price of electricity. For large systems, their cost is prohibitive.

Several proposals for multimegawatt solar power systems would use some of the electricity generated at peak periods to pump water up to a high reservoir. At night, the water would be allowed to fall through turbines, recapturing most of the stored energy. This is a costly solution, and it may not be practical for most solar energy systems, which are contemplated most often for flat desert terrain, where the sun is most abundant but potential reservoir sites can be many miles away.

The real answer is to find some way of storing the captured energy, just as oil retains the energy of plants long dead. We need a new fuel, one that does not contribute to global warming and other pollution problems. We have examined one such possibility already in our discussion of future transportation: hydrogen. It burns cleanly and is readily available, but it is expensive and requires a whole new distribution system. Fortunately, it is not the only option, particularly for large, fixed power systems.

Another possibility comes from the work of Dr. Moshe Levy, of Israel's Weizmann Institute. He uses solar heat to combine methane with either steam or carbon dioxide at 900 degrees Centigrade. The result is called synthesis gas. The gas retains virtually all of the solar energy that goes into the system. When the energy is needed, a catalyst breaks down the synthesis gas, releasing the heat and the raw materials, ready for reuse. In theory, the process could be 100 percent efficient. In practice, Dr. Levy has managed to retrieve up to 80 percent of the useful energy.

Synthesis gas has two main advantages: Unlike hydrogen, it can be distributed with the same equipment as compressed natural gas. Unlike natural gas, it releases no extra carbon dioxide when its energy is released.

We believe the hydrogen economy is a long way off. Synthesis gas is a much more likely prospect. We do not believe it will be widely used by 2010, not to power automobiles or to meet other general-purpose energy demands. However, it could well appear in future solar power stations as a convenient way to store energy for nighttime use.

GROWING POWER

The other way to collect solar energy for later use is nature's method: photosynthesis. Grow plants, harvest them, store them, and then burn them, or a fuel derived from them, when you need their stored energy. This is humanity's oldest source of energy. On the global scale, it may also be one of our newest. But there is more to "biomass" energy than throwing another log on the fire. It is not yet clear that plant-based fuels are a practical option for the developed world. However, most of the remaining questions will be answered in the next fifteen years.

After the trees that sustain many of the developing lands, most biomass energy today comes from agricultural and industrial wastes. In the United States, paper manufacturers generate nearly half of the energy needed to run their factories by burning sawdust, scrap wood, and waste from the pulping process. Denmark burns roughly 1 million tons of straw annually. Straw, the country's largest agricultural waste product, fuels small heating systems on some 12,000 farms and provides about 7 percent of Denmark's energy needs. In other countries, corn and wheat stalks, bagasse (sugarcane residue), and many other such waste products are available; even the dung from livestock can be converted to energy. David Hall, a biologist at King's College, London, estimates that if we could collect and use all the biomass now wasted by agriculture and forestry, it could meet 7.5 percent of the world's energy needs.

That would help, but it will not happen. Large agribusinesses can collect these resources efficiently and use them to provide some of their own power. But much of the world's agricultural waste is scattered across small farms in the Third World. In these areas, generating significant quantities of energy from waste is much less practical.

If biomass is to make a major impact on the world's energy economy, we will have to grow energy crops as we grow hay, but for burning rather than fodder. This should be cost-competitive with conventional energy sources. Energy from oil now costs about $3.15 per million BTU; energy from biomass would cost about the same, according to current estimates. But energy farming requires solving several problems.

One is identifying the most productive crops. An ideal energy plant grows quickly, stores energy efficiently, and grows back from the stump, so that people do not have to waste time—and energy—replanting after every harvest. Scientists at Oak Ridge National Laboratory have spent more than fifteen years evaluating potential energy crops for the United States. More than 100 woody species and 25 grassy species have been examined; six woody plants and one grass remain under study. Two seem to have clear advantages: poplar trees, especially hybrid varieties, and switchgrass. One or the other will grow almost everywhere that biomass farming is practical—everywhere, that is, except in arid and mountainous regions, such as those of the American West, and in tropic lands, where eucalyptus appears more suitable. Since 1980, genetic engineering and breeding programs have improved their biomass yields by some 50 percent. Some 62,000 acres (25,000 hectares) of poplars already are grown in the Pacific Northwest for use as paper and energy.

It is important to pick the best crops because planting them takes so much room. Photosynthesis is a clean way to collect energy, but it is a long way from being efficient. Plants capture only 1 percent of the energy that falls on them under the best of conditions; in the real world, their output usually is less than that. This means that it takes a lot of cropland to produce any significant amount of energy.

For example, the United States used a bit more than 82 quads in 1993; one quad equals 292 billion kilowatt-hours, or enough to run an average personal computer for something like 75 million years. It would take 150 million acres (60 million hectares) of land to produce 19 quads of energy per year, or a bit less than one-fourth of the current U.S. energy demand. And 150 million acres is about one-third of all the cropland in America. That statistic assumes that energy crops grow at a high average yield per acre; in the real world, even more land could be needed.

There are environmental questions as well. Energy crops would displace fossil fuels—at least, that is the goal—and that should reduce our net release of carbon dioxide into the air. In theory, energy farms this year would reabsorb all the CO_2 emitted by burning last year's crop. Unlike petroleum, they burn without releasing sulfur compounds. (Nitrogen oxides, unfortunately, emerge from almost any fire.) Energy farms could help to stabilize fields prone to erosion, another plus. For several reasons, agronomists believe that energy crops would require less fertilizer, water, and tillage than traditional farm plants, and these too are all net gains. There is some hope that energy crops would even sustain more diverse ecosystems than other farms. The problem is that energy farming would not merely replace traditional one-crop agriculture. To make a significant dent in our energy needs requires so much space that farms would have to displace rangeland and other relatively natural habitats. And that idea is not nearly so inviting.

Finally, at least for this discussion, there is the problem of how to extract the energy from the biomass. Burning wood or grass where it is grown makes sense; shipping it to some distant power plant does not. And the United States has a lot more need for energy in the populous coastal regions and in the industrial Midwest than it does in the central farmbelt, where biomass likely would be grown. So all that vegetable matter must be converted into some more convenient form. There are several obvious possibilities.

Under the right conditions, plants can be gassified. It takes heat, sometimes pressure and some other resources, depending on the exact process being used. But the gas produced burns cleanly and very efficiently. A gas-fired power plant captures about 80 percent of the energy in the original wood, almost twice as much as simply burning the plants themselves. Small gassifier generators are the most efficient of all, so this option is particularly good for providing local power in farming regions. A state-of-the-art generator in Värnamo, Sweden, uses gassified wood to spin a turbine engine. It produces 6 megawatts of electricity and 9 megawatts of heat for the surrounding community. Similar units could provide power for many towns throughout the world's farming regions.

Almost any once-living material releases methane when it decays. Wood, the tissue of grassy plants, and animal waste all can be used to feed anaerobic digestors and so be turned to methane. The process even leaves as its waste product a nutrient-rich sludge that can be used as fertilizer. The rotting garbage of landfills emits large quantities of methane, a substance that, molecule for molecule, contributes eleven times as much heat to the greenhouse effect as carbon dioxide. In Germany and Great Britain, many landfills have already been equipped to capture this gas and burn it for energy. There is no compelling reason not to convert biomass to methane.

The other obvious possibilities for a nonpolluting fuel are methanol and ethanol. We know already that alcohols make good fuels. Ethyl alcohol is widely used in the United States as a fuel additive, and one-third of the cars in Brazil run on pure ethanol distilled from sugarcane. But although the cost of alcohols has come down dramatically over the last decade or so, they still are not cheap. Enough ethanol to replace a gallon of gasoline costs about $1.50 (in 1994 dollars) in the United States at wholesale prices; gasoline averages less than $0.50 per gallon wholesale. No doubt further research will bring the price of ethanol, and methanol, down still farther. Whether it will drop far enough to allow competition with petroleum remains unclear.

In the long run, one study by the U.S. Department of Energy envisions an array of district powerplants scattered throughout the nation's farming regions. Each would supply power for the surrounding communities. Whether producing electricity or ethanol, each would process 1,000 to 2,000 dry metric tons of biomass (1,100 to 2,200 short tons) each day. Keeping them supplied would mean planting energy crops on 4 to 8 percent of the land within a radius of 25 miles (40 kilometers). Considering the acreage covered, that would make the local energy harvest the third or fourth largest crop in most areas.

In 1992, the United Nations commissioned a major study of biomass energy. The results were promising. By 2050, it forecast, fuel crops could meet no less than 55 percent of the world's energy needs. We doubt that seriously, if only because of the enormous land area required by such an ambitious biomass program. It certainly will not happen by 2010.

For the next fifteen years, biomass will remain a special-purpose energy source. Where agricultural wastes are readily available, they will be used for fuel. Large landfills will be tapped for their methane. One-fourth of the landfills in western Germany already are used to generate power; the United States and Japan will follow that example. But in 2010, scientists will still be trying to work out the details required to make biomass energy practical on the global scale. The amount of energy harvested from agricultural wastes and biomass farms will easily double or triple in the next decade or so. Yet that will account for only a small fraction of the world's total energy needs. In 2010 as today, most biomass energy still will radiate from small wood fires in the isolated regions of the preindustrial nations.

THE OTHER ATOMIC POWER

In the long run, probably the very long run, the real answer to our need for power is not to capture the energy of the sun but to manufacture it ourselves. This is the energy of the atomic nucleus. But instead of releasing it by breaking atoms apart, as occurs in today's fission-based reactors, the sun does it by pressing atoms together, by nuclear fusion.

As a concept, fusion power is simple. Two atoms contain slightly more energy than a single atom with the same total number of protons in its nucleus. So when the nuclei of two atoms fuse, they release their extra energy. Some of it takes the form of neutrons, the rest appears as electrons; some fusion reactions also release a few protons. That energy can be captured and used. It represents an inexhaustible source of power.

Fusion is inviting for reasons beyond the sheer size of the resource it offers. The necessary materials for fusion are abundant. Deuterium and tritium are the atoms that fuse most easily, two isotopes of hydrogen that contain one or two extra neutrons, respectively. Deuterium is found in nature, while tritium is easy to make. Fusion also is relatively clean, compared with nuclear fission. The end products of fusion are stable nuclei; they are not radioactive. The only nuclear waste that comes from fusion is the lining of the reactor itself, which is bombarded with neutrons throughout its working life. This is nasty stuff, but at least fusion reactors will produce much less of it than fission reactors do of spent nuclear fuel. Further, we will need to cope with it only after the reactor wears out; it will not accumulate in "temporary" storage facilities while politicians argue about where to put it.

Unfortunately, according to conventional physics, the only way to fuse atoms is by duplicating the incredible density of atomic nuclei that occurs naturally in the heart of the sun and then holding those particles together long enough for them to react. This means creating temperatures and pressures many orders of magnitude beyond any found here on Earth. Physicists have been trying to do that since the 1950s. Despite major successes, they still have not been able to "ignite" a fusion reaction and keep it going long enough to extract usable power from it.

There are two ways to create the conditions fusion requires. The older, and to date the more successful, is to confine a hot, electrified mixture of deuterium and tritium gas inside a "cell" of interlocking magnetic fields. The tokamak, the most effective design to date, consists of a ring-shaped chamber surrounded by two sets of electromagnets. One set of coils is a group of smaller rings wrapped around the tube through the hole in the doughnut. The other consists of two large rings, one above the doughnut and one below. Between them, these magnets create a spiral-shaped magnetic field around the chamber. They can squeeze the plasma inside densely enough for fusion and hold it there for about a second.

The second step is to heat the plasma. One way is to beam radio waves into the hot gas, heating it as a microwave oven pops corn. Another is to squirt more hot deuterium or tritium nuclei into the chamber. One such experiment with the tokamak at Princeton University produced temperatures and pressures high enough for a practical fusion-powered electric generator. In the half-second that it lasted, the reaction released more than 10 megawatts of power, enough to supply roughly 10,000 average American households for the brief moments the energy was available.

The second way to cause fusion is newer and less well proved but very promising. Instead of using a large chamber, so-called inertial confinement holds deuterium and tritium in a tiny capsule. To set off the reaction, the capsule is held at the center of a chamber and blasted from all sides with the most powerful lasers now available. (They grew out of research for the American "Star Wars" program and originally were intended to knock ballistic missiles out of the sky.) The capsule wall is vaporized, exploding outward and crushing the gases inside to densities more than a trillion times those achieved in the tokamak. The pressure lasts only one 10-billionth of a second, but that is long enough for fusion.

There is, or at least may be, a third way to cause fusion: cold fusion. If it really works, it could be the ultimate energy breakthrough and a revolution in physics. So far, no one is sure, except for the many skeptics who deny the possibility.

In 1989, Dr. Stanley Pons and Martin Fleischmann announced what sounded like an amazing discovery. Working at the University of Utah, they had found a new way to fuse deuterium. It did not require the heat and pressure that conventional physics said were irreplaceable. The two chemists simply placed two electrodes, one made of platinum, the other palladium, into a glass chamber of heavy water—water in which one of the hydrogen atoms has been replaced by its heavier twin, deuterium. Run a current between the electrodes, and the chamber gets warmer than it should; the heat energy emitted is greater than the electrical energy supplied to the electrodes. In many of their experiments, they found tritium, the product of deuterium fusion, and neutrons characteristic of the fusion reaction. Their results, Pons and Fleischmann believed, could be explained only by fusion of deuterium atoms that presumably were crowding together at the palladium electrode. They called the effect "cold fusion."

Most physicists had another explanation: Two chemists had been playing at matters beyond their skills and had bungled their research. Many hold to that view even today. Yet over the next few years, strange things began to happen in laboratories around the world. Many physicists reported that they had tried to duplicate the Utah experiments and had seen no evidence of fusion. Others found that the reaction did seem to release a little excess heat. Some thought they saw tritium. Some found helium, another product of fusion. Some reported neutrons. Many found one bit of evidence but not the others. The results of cold fusion experiments seemed exquisitely sensitive to conditions. A trace of normal water, the wrong kind of palladium, heat detectors that were not quite sensitive enough, even vibration—almost anything could stop the reaction or cause it to be overlooked. But it began to seem that Pons and Fleischmann had discovered something that science had not seen before.

We ourselves have mixed feelings about this kind of controversy. Though conservative by habit and nature, we cannot help remembering the words of Arthur C. Clarke, who was a respected physicist before he became a writer of science and science fiction. In a book titled *Profiles of the Future* (Harper & Row, 1963), in a chapter discussing failures of the imagination, he promulgated Clarke's Law: "When a distinguished but elderly scientist states that something is possible, he is almost certainly right. When he states that something is impossible, he is very probably wrong." Lest anyone misunderstand, he added, "Perhaps the adjective 'elderly' requires definition. In physics, mathematics, and astronautics, it means over thirty; in the other disciplines, senile decay is sometimes postponed to the forties."

Bearing that in mind, we have been inclined to give cold fusion the benefit of the doubt. It might well be a previously unknown chemical reaction rather than nuclear fusion; in fact, that explanation appears most likely. The chances that it would prove to be a practical new source of energy seem remote. But we have waited for the explanation with interest.

Now we are not sure what to think. Over the last year or two, research in the field has gone ominously quiet. At its 1995 meeting, the American Academy for the Advancement of Science arranged a conference on cold fusion. Almost none of the scheduled speakers showed up. No one has published a convincing obituary, but it begins to look as if cold fusion is dead.

Hot fusion, on the other hand, remains alive and well. Within the next few years, major new experiments are scheduled for both magnetic and inertial confinement. In 1996, scientists in both the United States and France hoped to start work on new laser facilities that should operate efficiently enough to release more fusion power than they use to start the reaction. This "scientific break-even" is considered a major goal en route to practical fusion power. And by 2000, a program called the International Thermonuclear Experimental Reactor (ITER) should complete the design of a new and much larger tokamak. If all goes well, early in the next century it should be able to sustain fusion reactions for up to an hour at a time.

Both the American laser facility and ITER depend for their funding in large part on the U.S. Congress, an institution not recently noted for its foresight. So these programs will probably be delayed.

Yet by 2010, we will know far more about the practical use of fusion power. ITER will be up and running, and it will produce enough energy to form the heart of a proof-of-principle fusion power plant. Inertial confinement fusion should be about to reach its next major milestone, yielding significantly more energy than is required to ignite fusion. It will take several more decades to bring the first fusion power station on-line. But fifteen years from now, we should know for sure that the technology will be there when we need it, when the oil finally does run out.

NURSING AN INJURED PLANET

During the nineteenth and twentieth centuries, an adolescent humanity has finally outgrown its planet. Our need for resources has approached the limits of Earth's capacity to sustain both our growing numbers and our immature technologies. That much, at least, is clear. We will spend the next two decades, and probably many years longer, working to understand the problems we have caused and figuring out what to do about them.

Until recently, the skeptics among us could reasonably doubt the substance of many environmental problems. Computer models of global warming differ widely in their predictions, and most forecast a much greater change in planetary temperatures than has occurred here in the real world. Ozone holes afflicted only the least populated regions of the planet, and when scientists bravely forecast the appearance of a hole over the Arctic in 1993, it failed to appear. The garbage crisis predicted a decade ago never materialized; small, inefficient dumps closed as expected, but larger, better-run landfills have taken up the load. Given these and other proofs that our knowledge of the environment is painfully incomplete, conservative scientists have been reluctant to draw conclusions about hazards that others have felt required an immediate response. Often the urge to procrastinate has received strong political support from industries likely to be inconvenienced by pollution controls and other such measures.

Yet in the past decade, many environmental problems have become visible to all. In the former Soviet Union and Eastern Europe, vast areas once closed to outside scrutiny have proved to be Dantean nightmares of toxic waste, spilled oil, and poisoned land, air, and water. In the United States, the Ogallala aquifer has been so depleted that farmers who once depended on it for irrigation have returned to dry-land agriculture. In Europe, the northern ozone hole has finally appeared, and wintertime levels of ultraviolet light are rising to what may be dangerous heights. And throughout the world, nuclear waste continues to pile up in short-term storage facilities while scientists and politicians argue over how to dispose of it safely. These and other environmental problems can no longer be denied.

During these next two decades, and probably for many years to come, damage to the world's environment will grow rapidly. So will our demand for resources, some of which already are being overused. We have built up too much economic and technological momentum since the Industrial Revolution to permit any quick or easy change.

Yet our understanding of these global injuries is growing as well, and our attempts to prevent and heal them are becoming more effective. By 2010 or so, we should have answered many of the remaining questions about what we have done to the environment. We may even have achieved our first successes in repairing the damage we once caused.

No single chapter offers enough room to examine all the proved and theoretical environmental problems the world now appears to face. What follows is a brief look at a select few of the issues most likely to require attention in the next fifteen or twenty years.

HARVEST OF THE HAMMER AND SICKLE

In Magnitogorsk, some 900 miles (1,000 kilometers or so) east of Moscow, children are dying. More than 500 have bronchial asthma and similar disorders severe enough to require hospitalization. Perhaps two-thirds of the city's children have chronic respiratory ills. Fewer than 1 percent are truly healthy, according to an estimate by local health officials. The cause is the giant Lenin Steel Works, 16 square miles of antique open-hearth smelters that employ 64,000 workers and spew out 650,000 tons (590,000 metric tons) of industrial waste every year. Emissions from the plant contaminate an area of 4,000 square miles (5,400 square kilometers). It is a scene repeated, usually on smaller scale, wherever the Soviet Union devoted itself to 1930s-style heavy industry.

Sovietologist Murray Feshbach and writer Alfred Friendly, Jr., summed up the environmental legacy of the Soviet Union in *Ecocide in the U.S.S.R.* (Basic Books, 1992). "When historians finally conduct an autopsy on Soviet communism, they may reach the verdict of death by ecocide," the two wrote.

> No other great industrial civilization so systematically and so long poisoned its air, land, water, and people. None so loudly proclaiming its efforts to improve public health and protect nature so degraded both. And no advanced society faced such a bleak political and economic reckoning with so few resources to invest toward recovery.

The situation has not changed much since Feshbach and Friendly examined it in 1992. The former Soviet Union and its onetime client states remain object lessons in what can happen when industrial and military goals dominate all other concerns. Merely beginning to heal these wounds will be one of the most important and difficult goals of the next twenty years. In most of these lands, it may not be possible.

Even a brief catalog of the devastation makes it clear just how difficult it will be to clean up after Soviet-style heavy industry:

- While the West agonized over what to do with its nuclear waste, the Soviet Union found an easy answer: It dumped tons of obsolete nuclear reactors and other radioactive material into the Barents Sea off the islands of Novaya Zemlya. Post-Soviet Russia continues to jettison reactor waste into the Sea of Japan, despite a promise to Japan that the practice would end. (In fairness, this is not a uniquely Soviet evil, merely one more case in which Russia retains the methods and attitudes of an earlier era. Between 1946 and 1970, according to a report delivered to the U.S. Senate Intelligence Committee, the Atomic Energy Commission sank at sea approximately 107,000 barrels of low-level nuclear waste, some of them only 12 miles off Boston.)

- Norilsk Nickel, which occupies a former prison camp built to supply it with labor, is one of the world's largest metal producers. Lying some 200 miles north of the Arctic Circle, it is surrounded for miles around by mounds of mine tailings, scrap metal, and toxic waste. But from an environmentalist's viewpoint, its most important product is air pollution. Its output of sulfur dioxide is nearly 3 million metric tons each year, some six times the emission from all the nonferrous metal producers in the United States. The plume can be detected as far away as Canada. Other mining centers are equally dirty. Several years ago, the Russian Environment Ministry estimated that one in every five tons of metal extracted from the ground was simply dumped into landfills.

- According to one estimate, one in every ten gallons of oil produced in Russia is spilled onto the ground or into the country's rivers. Foreign oil companies now rushing to develop fields once closed to them are modernizing the country's archaic wells and pipelines, but the work will take years to complete. Meanwhile, the losses continue. In October 1994, the third-largest oil spill in history covered the tundra near Osinsk with a pool said to be 7 miles (11 kilometers) long, 40 feet (12 meters) wide, and 3 feet deep (just less than 1 meter). Another pool in Siberia was 7 miles (11 kilometers) long, 6 feet (2 meters) deep, and 4 miles (nearly 5 kilometers) wide. A Western oil company has since scavenged the free petroleum, but the soil remains saturated.

◆ The Aral Sea is dying, and the effects of its demise are being felt around the world. Over the decades of Soviet rule, so much water was diverted from the Aral that too little is left to replace that lost to evaporation. As the Aral dries up, the salt and dust released have added 5 percent more particulate matter to Earth's atmosphere.

◆ The Baltic Sea has become a dumping ground for raw sewage, radioactive waste, heavy metals, and industrial waste of all kinds. More than one-fourth of the once-fertile sea bottom is dead, and the fish once caught there now are inedible. Experts from Finland, whose coasts are bathed in effluent from the Baltic, estimate that it will cost at least $1 billion per year for the next twenty years to clean up the mess.

◆ The sturgeon that once made the Volga famous for their caviar have almost vanished—and no wonder. An estimated 2.5 cubic miles (6.5 cubic kilometers) of sewage and industrial waste flow into the once-fertile river each year. That mass includes carcinogenic solvents, toxic aniline dyes, mercury and other heavy metals, petrochemicals, and a host of other noxious materials from some 3,000 factories, refineries, and nuclear power plants. In addition, a series of hydroelectric dams makes the river impassable for spawning fish.

On a lesser scale, similar problems afflict the Soviet Union's onetime client states in Eastern Europe. They are best understood in Germany, where state and federal governments have found massive environmental damage in the East. The Soviet military left behind 1,026 former encampments, most contaminated with spilled fuel, heavy metals, TNT, and even toxins leaked from chemical weapons. At a single former Soviet air base—Jüterborg, about 50 miles (64 kilometers) south of Berlin—investigators found no fewer than 9,000 patches of soil soaked with aviation fuel and the remains of munitions. Add to this toll the damage caused by decades of Soviet-style industry. Most of the electricity in East Germany came from high-sulfur brown coal supplied by open-pit mines next to the power plants; the generators poured unprocessed soot and sulfur directly into the air. Manufacturing followed the Magnitogorsk model, with no attempt to avoid polluting the air, water, and soil for miles around. Similar stories are found in Hungary, Poland, the Czech and Slovak Republics, and what is left of the former Yugoslavia.

These problems are every bit as daunting as they sound. Shortly after the reunification in late 1990, German authorities estimated that it would cost $140 billion to clean up the mess made by electric power plants alone. Restoring the former Soviet military bases, they said, could take 100 years. In Russia itself, the obstacles are greater still. Pollution there is caused largely by factories and power plants so outdated that they cannot be fixed but must be shut down. Yet these antiquated facilities remain essential to what is left of the Russian economy. Among its own workers and those of some eighty suppliers, the Lenin Steel Works alone provides the only available livelihood for an estimated 500,000 people. Next to that priority, pollution is unlikely to worry rulers schooled in the Soviet system, who in any case tend to view concern for the environment as an effete Western luxury. Long beyond the next twenty years, Russia and Eastern Europe will be struggling, either to heal their damaged environments or to live with them as they are.

AIR REPAIR

Two of the more deniable environmental problems afflict the air on which our survival depends. Global warming, a few doubters still maintain, may not result from human actions but from a natural process that is unlikely to threaten serious consequences. And while ozone depletion by chlorofluorocarbons (CFC) has become all but impossible to deny, its effects are nebulous at best. Those ideological positions are wishful thinking, and they are growing more difficult to defend each day. Fifteen years from now, even the most committed skeptics will have been silenced. Science at last will understand the atmosphere well enough to spell out the origins and consequences of greenhouse warming and ozone depletion. It may even have devised practical ways to correct these problems. We are surprisingly close to these goals today.

Until recently, global warming has been remarkably difficult to pin down. Yet the three essential facts are beyond argument:

- ◆ Climatologists have known about the greenhouse effect for well over a century. It occurs in nature. Carbon dioxide, methane, and other components of the air retain some of the sun's energy, which helps to warm the planet. If not for that, Earth's surface would average about 16° Celsius (C) below freezing instead of 17° C above.

- ◆ The level of carbon dioxide and other greenhouse gases has been rising since humanity first began to burn large quantities of coal, some 250 years ago.

- ◆ At the same time, Earth has been growing warmer. Early records are incomplete, but it appears that the planet's average temperature rose by about ¼° C over the 100 years beginning in 1780. It climbed another degree the next seventy years, until 1950. Since then, it has risen by roughly another ½° C. The warming trend that ended the last Ice Age amounted to little more than 5°.

All this seems to make a pattern, and you already know how the argument goes from there: We have poured so much pollution into the atmosphere that we have raised the temperature of the entire planet. If we continue to fill the air with carbon dioxide, methane, nitrogen oxides, CFCs, and other heat-trapping gasses, this warming will continue. Eventually, it could change climates throughout the world. In the United

States, the grain fields of Kansas and Nebraska could become semitropical. Britain would see the last of its fabled fogs—and if the polar ice caps melt, the world could see the last of much of Britain. The Maldives and other inhabited islands would vanish beneath the waves.

Of course, the situation has proved to be more complicated than that. The warming seen over the last century or so turns out to be only about half of what would be expected if greenhouse gases were the only influence now affecting the world's temperature. At the least, some other factor must be slowing the warming process. There are many possible candidates. Scientists still know too little about the formation and effects of clouds and atmospheric dust, both of which reduce the solar energy that reaches the planet's surface. There even is a small chance that the current warming trend has nothing to do with the greenhouse effect. Earth's climate has warmed and cooled often over the last few million years without human help. It could be changing again for purely natural reasons that we do not yet understand.

However, two recent studies make that seem much less likely. Both involve new, more accurate computer models of Earth's climate. Both examined the effect of a second kind of pollution emitted by industry, sulfate compounds. Like carbon dioxide, sulfates are released in the burning of fossil fuels. They are best known as the cause of acid rain, but they also form particles in the air that reflect part of the sun's energy back into space. According to scientists at Lawrence Livermore Laboratory, in California, and at the British Meteorological Office, these sulfate aerosols are slowing the greenhouse effect. Plug them into climate models, and the expected rate of global warming turns out to be almost exactly what has occurred in the real world. Statistically, there is little chance that natural processes would create such a precise match.

This research offers the best evidence yet that air pollution is damaging Earth's climate. According to the British study, greenhouse warming will add roughly another 1.25° C to the planet's average temperature by the year 2050.

Thanks to this research, we believe that for practical purposes, the long-standing argument about global warming has been settled. Over the next twenty years, further study will fill in many of the details that scientists do not yet understand. It will not invalidate the basic insight: By polluting the air, humanity has begun to change the world's climate in ways that may prove destructive, and even dangerous.

Ozone depletion is still better confirmed. Over the past ten years, scientists have come to understand it quite well. Released into the air, CFCs are almost indestructible until they make their way into the upper atmosphere. Once there, ultraviolet light (UV) breaks them down, releasing chlorine. By a complex series of chemical reactions, a single molecule of chlorine can destroy as many as 100,000 molecules of ozone. Each spring, when atmospheric conditions are right, these reactions destroy virtually all the ozone in a circular patch of sky over the South Pole. That patch grows larger each year. In 1995, for the first time, an "ozone hole" appeared over the Arctic as well.

Where the ozone thins, more ultraviolet reaches the ground. Even at normal levels, UV causes sunburn, cataracts, and skin cancer. We can expect these problems to become more common and severe as the ozone layer gets thinner. Some scientists believe that excess UV is already damaging ecosystems in the Antarctic. Over populated areas, the least it could do would be to raise cancer rates. In 1995, the northern ozone hole raised the amount of UV bombarding populated sections of Scandinavia.

Skeptics have argued that ozone depletion will never cause serious harm; after all, the holes appeared only near the poles, where human life is spread even more thinly than the ozone overhead. Unfortunately, this is like suggesting that a bathtub will never empty because water disappears only at the drain. Each year, more ozone is destroyed over the Antarctic than is created above the entire world. Throughout the year, ozone from other regions diffuses toward the poles, where it breaks down the following spring. Since the effect was discovered in 1985, the planet's supply of ozone has declined by about 6 percent, and the amount of UV reaching the ground has increased fractionally. Because CFCs persist in the atmosphere for up to a century, we can be sure that this decline will continue for decades to come.

What to do about air pollution is another question, and one that is being hotly debated. Working Group 3 of the United Nations Panel on Climate Change is charged with weighing the effects of global warming. At the panel's meeting in December 1995, the group's original report concluded that nothing should be done to halt greenhouse warming. The damage likely to be caused by climatic changes, it held, will cost somewhere between 1.5 and 2 percent of the gross world product (GWP); effective remedies would cost significantly more than that. Other estimates have put the cost of damage at anything from 12 to more than 100 percent of the GWP, which would make antipollution measures cheap at almost any possible price. The group's final report was changed for largely political reasons, and the debate continues with little sign of resolution.

How much would it cost to stop global warming? Several years ago, the National Academy of Sciences (NAS) studied the problem for the U.S. Congress. Its report examined fifty-nine so-called mitigation strategies designed to reduce emissions of carbon dioxide and other greenhouse gases. The findings were not encouraging. For example, making cars more efficient costs from $1 to $15 per ton of CO_2 saved, and that is one of the cheaper options available. The tab for other proposals ranged up to several hundred times higher. Use all the relatively cheap options—say, under $10 per ton of carbon dioxide saved, or the greenhouse equivalent in other gases—and they would reduce global warming by 20 to 40 percent. In all, the potential bill for a stable climate came to at least $500 billion per year.

But there were some "wild-card" options as well, mitigation strategies that would deliberately change the climate to offset global warming. Most of these so-called geoengineering concepts lay at the very borders of the scientifically credible; some were well beyond. Many might not work at all. Yet they offered at least the hope that we can effectively combat global warming at a price that humanity can afford.

One of the most spectacular ideas involved orbiting giant mirrors in space to reflect some of the sun's warmth away from Earth. If each mirror were 100 million square meters—about 39 square miles—it would take 55,000 of them to counteract the world's output of greenhouse gases. Offsetting just the American contribution to global warming would require 110 mirrors. But the cost of lofting them on the Space Shuttle would reach at least $120 billion, not counting the reflectors themselves. It's a price likely to dissuade even Washington bureaucrats.

One obvious alternative is a "space parasol," a dust cloud to filter out a bit of solar energy. But if the dust particles are small, so that it would take only a few tons to cover the required area, keeping launch costs to a minimum, the solar wind would quickly sweep them back to Earth. And if the particles are large enough to remain in orbit, launching them would be prohibitively expensive. Eliminate this notion as well.

The most promising way to get rid of excess CO_2 originated with the work of the late John Martin, of California's Moss Landing Marine Laboratory. In widely separated parts of the Pacific, biologists knew, the water is rich in nutrients like nitrogen and phosphorous, yet microscopic plants grow poorly. This odd sterility occurs both near Antarctica and around the equator. Martin set out to find what limits the growth of these phytoplankton.

The puzzle was important for global warming because marine algae absorb carbon dioxide. In fact, they provide one of the most important natural mechanisms for getting rid of excess CO_2. Small, shrimplike creatures called krill eat the algae and deposit the carbon in their fecal pellets, which fall more or less permanently to the ocean's floor. Grow enough algae, and the greenhouse effect should disappear with the krills' waste. This is the only geoengineering scheme yet tested on anything like the scale required for practical air repair.

Dr. Martin succeeded in learning what limited the growth of algae in the barren areas of the Pacific. It turned out to be iron. The problem is that almost all the iron in the sea arrives in dust blown from land, and the barren areas are so remote that little dust ever reaches them. Fertilize a beaker of water with iron, and the algae bloom.

The procedure could work on a planetary scale. In one practical test, it succeeded better than anyone dared hope. Dr. Kenneth Coale and his colleagues, also from Moss Landing, took a small ship to an area off the Galapagos islands and spread a half-ton of iron across 100 square kilometers (38 square miles) of barren water. Literally overnight, the clear blue water turned green as thirty to forty times more microscopic plants suddenly appeared. In the few days that the effect lasted, the algae absorbed something like 350,000 times more carbon dioxide than before.

Now think really big. According to the National Research Council, 270 ships could fertilize 18 million square miles (47 million square kilometers) of ocean. That could eliminate as much as one-fourth of the excess CO_2 from the atmosphere at a total cost of between $10 billion and $110 billion per year. It seems a reasonable price to pay.

Those ships might try burning sulfur as they steamed back and forth across the South Pacific. This would send an aerosol of sulfur dioxide into the lower atmosphere. In theory, water condensing on the aerosol droplets should create artificial clouds. And it would take only 4 percent more cloud cover over the oceans to offset the warming caused by a doubling of atmospheric CO_2. One back-of-the-envelope calculation puts the cost of eliminating the greenhouse effect at about $1 per ton of CO_2 and perhaps as little as $0.03 per ton.

Several other geoengineering proposals would attack ozone depletion. For example, ethane or propane should block a key link in the chain of chemical events that destroys the ozone. So use high-flying airplanes to dump 50,000 tons or so of the cheap hydrocarbons into the air over the Antarctic each winter. And when spring arrives in September, the ozone hole should fail to arrive with it.

Alternatively, CFCs can be broken down by infrared light from a powerful, carefully tuned laser. Just install the proper lasers atop the highest mountains and fire away. Unfortunately, today's lasers are not nearly efficient enough for the job.

But the best idea to date comes from physicist Alfred Wong. He suggests using a fleet of thirty to fifty blimps to drag electrically charged wires around the sky. These arrays, the size of a football field, would extract the chlorine from the air before it could do any damage. The process works in the small test chambers of Wong's laboratory at the University of California-Los Angeles. Other scientists have questioned whether the process can be expanded to the giant scale required to clean the atmosphere, but Wong himself has few doubts. If he is correct, the dirigibles required to correct ozone depletion would cost "only" $400 million or so. In the world of geoengineering, that is a bargain.

Nearly a dozen more geoengineering concepts have been floated to deal with global warming, but you get the idea. Cleaning up the atmosphere may be more than the analysts at the United Nations can cope with, but it clearly is not beyond the scientific imagination.

Each of the ideas proposed so far has disadvantages, many of which could rule them out on closer examination. Fertilizing the ocean seems benign at first glance, but all those krill feces would be eaten by bacteria. Those microbes, in turn, need oxygen. They might conceivably monopolize so much oxygen that other deep-ocean life would suffocate. Filling the air with dust or sulfur compounds might not work because they would block out the sun's heat only during the day, while greenhouse gases trap heat during the night as well. The planet's average temperature might come out right, but the temperature difference between night and day could still be changed, with unknowable results. Some other geoengineering concepts would avoid any obvious potential catastrophes yet remain beyond our current technical ability. Many would be too expensive. Still others appear technically and economically feasible, yet they could not be implemented on a scale large enough to eliminate more than a small fraction of the greenhouse gases or ozone-depleting CFCs that must be dealt with.

Yet if today's ideas prove unworkable, tomorrow's may well be better. With that in mind, we will make a very tentative forecast, little more than a guess of the kind we usually try to avoid: By 2010 or so, scientists will devise at least one workable geoengineering approach to combat global warming and one to protect the ozone layer. These ideas will not be put into effect by then, and for political and philosophical reasons they may never be used. But scientists and policy makers will be studying them vigorously. By that time it will be clear that answers to global warming and ozone depletion are urgently needed.

NUCLEAR FOLLIES

In 1990, the United States had stockpiled about 15,000 tons of nuclear waste, most of it in "temporary" storage at the power plants and nuclear arms factories that generated them. At that point, the federal government was scheduled to move it into a permanent storage facility by 1998. At the end of 1995, there were about 30,000 tons of radioactive waste waiting for a long-term resting place. The federal government still was legally required to take possession of it in 1998, but political pressures had delayed any possible opening of the government's proposed storage site until at least 2010. By 2010, something like 65,000 tons of nuclear waste will have accumulated. The temporary storage facilities now in use will have been filled long since. There is little cause to hope that permanent storage will be available even then.

The United States is far from being the only nation where political pressures have blocked anything but short-term approaches to the long-term problem of radwaste. Taiwan has three operating nuclear power plants, which produce roughly one-third of the nation's electricity. Its main storage site for spent fuel rods, worn-out reactor parts, contaminated protective gear, and the like is located on Orchid Island, off Taiwan's southeast coast. Waste material there is mixed with concrete, put into containers, and buried in trenches 10 feet (3 meters) deep. Since 1982, more than 96,000 containers have been buried on the island, and the facility is nearly full. Taipower, the state electric company, wants to add space for another 59,000 containers. The idea does not sit well with the island's 3,000 inhabitants. They are members of a tiny tribe known as the Yami, and they accuse Taiwan's government of racism and attempted genocide. The existing containers are corroding, they charge, and may leak into the water supply. They also claim that the waste already there has caused an increase in the rate of cancers and birth defects. Rather than welcoming more reactor waste, the Yami have threatened to dig up what has already been buried and throw it into the sea. Work on extending the site has been halted and may well have been abandoned permanently. That leaves room at the site for only 1,800 more containers. Taiwan has tentatively solved its problem by contracting to send its low-level nuclear waste to North Korea for disposal.

Japan has a waste problem as well. Rather than burying used-up fuel rods, it intends to reprocess them to recover the "unburned" atomic fuel. Unfortunately, Japan lacks its own reprocessing facility, and there is little chance that the government could build one against the overwhelming weight of public opposition. So it intends to send its nuclear waste out for reprocessing. The material will go by ship all the way to France and back.

England is in even a worse fix. American authorities at least are still trying to find somewhere to store their nuclear garbage. Their British counterparts have given up. After surveying some 500 possible sites, they had settled on a location near the Sellafield fuel reprocessing plant in the north of England. There they intended to build underground storage for low- and intermediate-level wastes. Construction of the repository was scheduled to begin in 2005. In 1995, however, opposition groups managed to scuttle the plan. Construction will not begin until at least the year 2060.

Yet in some ways the United States is the world's leader in avoiding the hard question of what to do with nuclear waste. It has been arguing the question, to no obvious effect, for decades.

Twenty years ago, the country had more potential storage sites than it could ever need. Many parts of the country are home to deep salt deposits, which are ideal for the permanent storage of almost anything you never want to see again. Geologically, they are all but inert. Most are in seismically quiet areas, and in any case they are not prone to the kind of cracking that might eventually expose something buried in rock strata. A nearby earthquake could shake a salt formation, yet anything buried deep within the salt itself would remain sealed off from the outside world. By definition, salt formations also are free of water that might corrode waste containers. If any significant moisture was available, the salt would have washed away millennia ago. The United States holds so many such formations that it could store all the nuclear waste it would produce in centuries of reliance on atomic power and still have space to rent to those in less fortunate lands. Yet over the years, one by one, political opposition managed to seal off the salt beds, leaving the Department of Energy to search for somewhere else to put its radwaste.

Thus far, we have spoken as if nuclear waste were one homogeneous material that required a single response. In fact, there are three varieties of radwaste, each of which should be considered separately.

High-level nuclear waste, the kind we usually think of, consists of the spent fuel from reactors plus liquid wastes from the manufacture of atomic weapons. It is intensely radioactive, and many of the elements within it have long half-lives. It is this material that must be hidden away virtually forever. Fortunately, it does not take much space.

TRU, or transuranic waste, includes anything contaminated with plutonium from weapons programs. It includes parts of the machines used to mill plutonium, protective clothing, and even soil contaminated with the metal. Compared with high-level waste, TRU is not particularly dangerous. The radiation emitted by plutonium can be stopped by a piece of paper. But plutonium is highly toxic—inhaled, it causes lung cancer—and its half-life is more than 24,000 years. It can be hidden behind a minimum of shielding, but it must remain hidden for a very long time.

Low-level waste is little more dangerous than ordinary trash. Most of the radioactive materials in this category are short-lived, like the isotopes used in medical laboratories. The protective gear, tools, and other debris from nuclear power plants may hold traces of long-lived radionuclides, but no more. American courts have ruled that low-level wastes can go into relatively ordinary landfills.

They don't. The fear of radiation is so powerful, and political opposition to dumping so strong, that the entire country has only two old landfills for the permanent storage of low-level waste, one in Barnwell, South Carolina, the other in Richland, Washington. They serve regional consortia of nineteen states. The rest of the country has nowhere to deposit even this safest of all nuclear materials.

California has been trying for nearly a decade to set up a third site near the tiny town of Needles, far off in the Mojave Desert. By all accounts, it would take a long and profoundly unlikely chain of mishaps for radioactive elements to find their way from the landfill to populated areas. Yet plans for the dump will remain blocked by litigation. The federal government, which owns the 1,000-acre (405-hectare) site of the proposed facility, now refuses to turn the land over to the state of California, which would operate the landfill. No resolution is expected until at least 1998.

Yet this is only a minor skirmish compared with the war against the country's largest and most important nuclear dumping site. Deprived by political pressure of its unique natural depositories, the salt beds, the U.S. Department of Energy went looking for somewhere else to store high-level nuclear waste. After several years, Congress stepped in and decreed that only one high-level facility would be developed, at Yucca Mountain, Nevada. Over the last ten years, the government has spent more than $1.5 billion to study the site's geology, tectonics, hydrology, and other factors that might affect its suitability for "file-and-forget" waste disposal. A host of theoretical and practical objections have been raised to the Yucca Mountain depository, including the discovery of three geologic faults that cross the site. And as geologists reckon things, volcanic activity is not that long removed from the area. The Yucca Mountain depository originally was scheduled to open in 1998. That date has slipped, first to 2003 and now to 2010. At the rate things are going, the British may open their Sellafield waste dump before Yucca Mountain sees its first ounce of nuclear material.

Wherever nuclear waste is mentioned, the story is the same. Environmental activists bring up a host of arguments against the site, many of them wildly unlikely but seldom entirely impossible. The proposal is tied up in litigation. Political pressure builds. And the problem remains unresolved. After nearly a decade in development, the Waste Isolation Pilot Plant, a TRU disposal facility at Carlsbad, New Mexico, remains unused. A temporary storage site in New Mexico is in the courts, despite the fact that the Mescalero Apache Tribe, which owns the land, has voted to accept the material. At least one state has shut down its low-level waste repository. Nuclear waste is one environmental issue that no one wants to touch.

This is not yet a crisis; there is no reason it should ever become one. The temporary-storage facilities at most nuclear power plants are perfectly capable of holding the waste until society settles on a long-term solution to the disposal problem. At least, they would be if they were allowed to do their job. However, many of those holding sites are growing full, and a few have begun to leak their radioactive contents into the surrounding soil. And every attempt to enlarge or improve them meets the same kind of legal and political opposition that has blocked Yucca Mountain. The threat of a serious leak at one of these on-site holding facilities grows with each delay.

One cynical theory holds that this represents a deliberate strategy by the antinuclear lobby. If the on-site holding facilities get full, the reasoning goes, the activists hope that reactors will be forced to shut down, and they will have arrived at their goal through the back door.

That truly would be a crisis, but it will not happen. Instead, eventually, the government will impose a new "temporary" storage site on whatever state is least able to defend itself in Congress. The order may not go out before 2010, but it will not be delayed much beyond that. While it will be an imperfect answer to the problem of nuclear waste, it may be the only solution available.

NOR ANY DROP TO DRINK

The vast Ogallala Aquifer is running dry. A huge pool of water that lies beneath the American Great Plains states, it has supplied most of the water that supports the fertile farmlands of Kansas, Nebraska, and parts of Colorado, Oklahoma, New Mexico, and Texas. The Dust Bowl of the 1930s is farmland today, because irrigation systems tapped the waters of the Ogallala and drew them to the surface. Once this aquifer contained as much water as Lake Huron. But each year, much more water is drained from it than rainfall puts back. Over the last four decades, the water table has dropped by well over 100 feet (30 meters). Farmers in Texas already have been forced back to dry-land agriculture. When nature fails to deliver rain, the Dust Bowl will return.

For Texas farmers, this is a harsh blow; yet it may be the least of the world's water problems. According to the World Bank, at least twenty countries already are short of water. They include much of the Middle East, plus parts of central Africa and such unexpected locales as Barbados and Singapore. Already Amman, the capital of Jordan, is forced to ration its water. In Taiz, Yemen, municipal water flows through the city's plumbing only once every three weeks.

In the Middle East, water is politics. Israel depends for water on the same complex of seas and aquifers that feed its neighbors, Lebanon, Syria, Jordan, Saudi Arabia, and Egypt. Of these parched lands, only Lebanon enjoys even a small surplus of water. The region suffers droughts every four years, on average. The Palestinians' most serious complaint against Israel does not involve political or military repression, but thirst. The 5.5 million Israelis use four times as much water per person as the 2 million Palestinians and in the past have forbidden the Palestinians to deepen their wells or to sink new ones into the aquifers on which both peoples depend.

Where water can be found, it is seldom clean. Around the world, something over 1 billion people are deprived of clean water, often because their own sewage flows into nearby rivers. In Latin America, only 2 percent of sewage is treated; few undeveloped lands do much better. According to most estimates, polluted water causes about 80 percent of the disease in the Third World, killing perhaps 10 million people each year.

Even in the developed lands, water cannot always be trusted. As recently as 1993, nearly half the residents of Milwaukee, Wisconsin, were sickened by a microbe that escaped the city's waste treatment plants. According to the National Resources Defense Council (NRDC), water-borne bacteria sicken perhaps 900,000 Americans each year and kill 900. Worse yet, the chlorine used to kill those pathogens produces a family of chemicals called trihalomethanes, which can cause bladder and rectal cancer—more than 10,000 cases each year, according to the NRDC. And despite the strict licensing requirements mandated under the Clean Water Act, no one really knows what chemicals industry discharges into the American water supply. One study of 236 companies found that 77 percent of the materials they released were not listed on their permits. According to the General Accounting Office, most of these chemicals "are recognized as human health risks."

For the next few decades, the growing shortage of drinkable water may be the world's most serious environmental crisis. By 2025, the Middle East's population will double. Its demand for water will quadruple. Iraq and Syria soon will lose much of the water they now receive from the Euphrates (60 percent and 40 percent, respectively); instead, the river's flow will be diverted into a huge Turkish irrigation project miles upstream. Elsewhere, Haiti, Cyprus, Zimbabwe, Tanzania, and Peru all will be running dry. At present, some 600 million people live in arid and semiarid lands; by the middle of the next century, nearly 5 billion people will inhabit regions where water is scarce.

If global warming materializes, the situation could be even worse. Areas of drought would spread. A small rise in sea levels would flood the Netherlands, the mouth of the Nile, Bangladesh, and other low-lying areas. Refugees from many such areas would add to populations where water is already scarce. Salt would contaminate the fresh water supplies of many other regions.

The obvious answer to water problems is desalinization. Just draw water from the sea, remove the salt, and drink. Desalinization requires so much energy that it seldom is practical, save in the oil-rich states of the Persian Gulf. Because the salt quickly clogs the membranes used to purify water, it also takes a lot of hand labor to keep a desalinization plant running. There is little prospect that new purification methods will ease the situation; this is a mature technology.

Instead of solving the problem of water shortages with a single "magic-bullet" approach, the world will adopt many smaller-scale answers, each tailored to local conditions. The developed nations already are making better use of their water. Even adjusting for inflation, Japan now squeezes almost four times as much industrial production from each gallon of water as it did in the 1960s. In the same period, the United States has nearly quadrupled its factory output while reducing its industrial water use by nearly one-third. California's Central Basin District, like many others around the world, is recycling treated wastewater for use in industry and the irrigation of park lands. Israel, Australia, and other lands use recycled wastewater to irrigate farmland. Agricultural regions with even less water to spare also are improving their water use, often by returning to conservation methods used by their ancestors. One well-known program in Burkina Faso built low stone walls around the contours of farmlands to slow the runoff of rainwater; crop yields have gone up by well over 50 percent. Other programs have achieved similar results by terracing hilly croplands, using grasses to retain what little water the land receives, or even setting aside large tracts to capture rainwater and funnel it to small garden plots.

Thus far, most of these programs have been remarkably low-key. For some reason, water problems have seldom drawn much attention outside the areas most immediately affected by them, save when droughts in Africa have put images of starving children on the television screens of America and Europe. That is unlikely to be true for much longer. Water problems cut across too many national and regional boundaries for them to be solved locally. Over the next twenty years, as water problems get worse, many international bodies will take up the cause of water conservation. This may be the best hope of survival for the people of the arid lands.

ONE-WORLD ENVIRONMENTALISM

This growing internationalism will affect most environmental problems in the coming decades. Global warming, ozone depletion, nuclear waste, water shortages, and even the compound ecological nightmares of the former Soviet Union clearly are international problems. So are most other environmental issues. Russia still produces more CFCs than any other nation, and there is some question whether its rigid, inefficient economy will be able to phase them out as the Soviet Union once agreed to do. China plans to power its industrial development with coal; its hundreds of new power plants and factories could add 20 percent to the world's output of carbon dioxide and increase the threat of global warming. Scientists in Europe have wondered for years why acid rain continues to kill the Continent's fish and trees, even though European sulfur emissions have dropped by 30 percent in the last fifteen years; they recently discovered that sulfate from the United States makes its way across the Atlantic to fall on lakes and forests that otherwise would be unpolluted.

International efforts to control environmental problems are still a relatively new idea. The Montreal Protocol to eliminate CFC production by the end of the century was the first major effort in this area and thus far the most successful. International agreements to fight global warming have been scuttled by the United States; both the Bush and the Clinton administrations concluded that limiting the release of carbon dioxide would be too costly to justify when scientific evidence of the problem could still be argued. In this, they defied a scientific and political consensus that virtually all other powers have accepted.

Over the next twenty years, this kind of cowboy individualism is likely to fade away. Already it is becoming inescapably clear that few environmental problems will wait until we find it convenient to deal with them. Thus the United States eventually will agree to limit its sulfate and carbon dioxide emissions. And the Montreal Protocol will turn out to be just the first of many international efforts to heal the injured planet on which our survival depends.

9

MEDICINE FOR
THE NEW MILLENNIUM

Until recently, it took some four decades for the discoveries of pure science to find their way into the everyday miracles of technology. The vast chemical industry of the 1920s grew from advances that organic chemists had made in the 1880s. The primitive atomic weapons of 1945 were based on work that Albert Einstein published in 1905. Even the "miracle" plastics of the 1960s and 1970s were built on chemical developments from the Depression era.

Today, that development cycle has been reduced to a few months, and even in the purest of sciences, many laboratory researchers can tell inquiring reporters or meddlesome funding managers what practical applications might eventually emerge from their theoretical efforts. Modern society's hard-headed demand for a short-term payoff—the modern equivalent of academia's traditional "publish or perish"—simply requires it.

Nowhere in science do new findings move from the laboratory to daily practice faster than in medicine. Almost by definition, biomedical research aims not to understand but to heal. And though a possible new treatment must work its way through a long, formalized series of studies to prove its safety and efficacy, when scientists announce the results of experiments in mice, we almost always know what new therapies to expect for human patients. And they will not take forty years to arrive but ten, or even five.

On that basis, this is one of the easiest forecasts we have ever made: The first decade of the twenty-first century will be one of the most remarkable and productive in the history of medicine—not just in history to date but in all the history there will ever be.

Dozens, even hundreds, of new treatments will reach beleaguered patients. Doctors will learn to repair damaged heart muscles, prevent Alzheimer's disease, avoid scarring after severe cuts and burns, wean addicts from drugs with relatively little hardship, and prescribe a host of new psychoactive medications capable of easing mental illness more effectively and with fewer side effects than any now available. They will build artificial replacements for failing organs. And, of course, they may develop a vaccine against HIV, the virus that causes AIDS.

All that would make the next fifteen years a banner period—but there is more. No fewer than four all-out revolutions will transform the practice of medicine in the next decade. These are far-reaching developments, the kind of discoveries that change both our understanding of entire fields and the way doctors approach their patients. Rather than trying to cover the many individual advances that will contribute incrementally to medicine over the next few years, we will focus on these few transcendent developments.

Our candidates for the next medical revolutions are:

- ◆ Gene therapy. In the last ten years, biomedical researchers have learned more about the workings of our genes than at any time since the discovery of DNA's famed "double helix," some forty years ago. In the next decade, that theoretical understanding will be applied as practical medicine. Doctors will learn not just to treat hereditary diseases but to cure and even prevent them. They may also reduce our susceptibility to disorders that, if not directly inherited, at least are influenced by predispositions that are governed by our genes.

- A cure for cancer. In the early 1970s, President Richard Nixon declared a $10 billion "war on cancer." Today, we understand this fearsome disease much better as a result of that research effort. Yet the sought-after miracle has remained out of reach. In the United States alone, more than half a million people die of cancer each year, while another 1.2 million new cases are diagnosed.

  Now, at last, the time may have come. Scientists, and one man in particular, appear to have figured out what it is all cancers have in common and why our bodies do not fight them off as they do other diseases. And based on that understanding, a treatment is obvious. Though much more testing remains to be done, it seems a true cure for cancer is it hand. We expect it to be in common use a decade hence.

- Hormone replacement therapy for the symptoms of aging. Women have used hormone replacement for years to ward off the sudden, highly visible aging that often follows the loss of their natural estrogen supply. Now this principle is being broadened to cover many of the other hormones whose levels decline as our years advance. As a result, we should retain more of our health and vigor far into old age and be much less susceptible to heart disease, cancer, and other age-related disorders. There even is strong evidence that supplements of one hormone, which ordinarily declines over time, may delay the aging process itself.

- Rejection-free transplants. Four decades after the first successful human heart transplant, doctors are still struggling with the same obstacle that made transplants so difficult in the first place: The patient's immune system recognizes the donated organ as foreign and sets out to destroy it. Soon the organ dies. The patient follows. Even with powerful drugs to prevent rejection, scarcely more than half of patients who receive new lungs are still alive a year later, while heart transplant specialists consider their work a success if the patient survives for just five years. Many potential recipients never get a transplant at all, because no compatible donor can be found in time.

Both those problems—rejection and the shortage of donors—would be solved if it were possible to implant foreign organs without provoking the patient's immune system. It appears that such a technique has been found.

Four revolutions in little more than a decade, each of which should cure tens of thousands of patients with disorders that lie far beyond the powers of today's medicine.

For perspective, consider that there have only been three genuine revolutions in medicine to date. The first occurred in the late seventeenth and early eighteenth century, when cities first banished sewage from their streets and sent rat-catchers out to wage war against vermin. Sanitation and other public health measures dramatically reduced the rate of death from infectious disease. The second was the discovery by Crawford W. Long in 1842 that ether could relieve the pain of surgery. Suddenly, surgery became a practical healing tool, not just a desperation measure, and for the first time doctors could actually heal a disease—by removing, say, an inflamed appendix—rather than simply treating the symptoms. The third revolution stretched from 1796, when Edward Jenner introduced the first vaccination for smallpox, through the discovery of penicillin and other antibiotics in the 1930s and 1940s. Thanks to this work, doctors today routinely prevent and cure infectious diseases that in the past killed millions. We believe that each of these new revolutions-in-the-making will turn out to be important enough to join this select group of healing miracles. They will improve our lives in ways that past generations could only dream about.

In this chapter also, we discuss one more development, an experimental operation called omental transposition. We do not view it as a revolution on the scale of the other developments we cover. Yet it is a remarkably versatile technique, and, though it is being studied energetically in other countries, it has been almost completely neglected in the United States, where it was developed. Consider this section a public service announcement on behalf of a needy and very deserving branch of medicine. It too could change the way doctors treat their patients in the next fifteen years.

NEW GENES TREAT AGE-OLD ILLS

In the revolution against inherited illness, the "shot heard 'round the world" was an infusion of white blood cells carrying synthetic DNA. It was delivered in September 1990 by Dr. W. French Anderson, then with the U.S. National Institutes of Health (NIH). The patient was a four-year-old girl named Ashanti DeSilva. She was the first human being ever to receive an artificial gene intended to cure an hereditary disorder.

As a species, we suffer from no fewer than 4,000 known ailments caused by defects in a single gene. These include some of the most familiar inherited diseases and some of the most destructive. Familial hypercholesterolemia, which is carried by one in 500 people of North European descent, is the most common of known genetic defects. When inherited from both parents, it results in fatally high cholesterol levels. It is caused by a mutation in a single gene. Cystic fibrosis, the second-most common genetic illness, fills the lungs with sticky mucus and slowly smothers its victims. It too stems from one pinpoint error in the DNA. Hemophilia, sickle-cell anemia, Huntington's disease, Tay-Sach's disease, and even some cases of bedwetting all are caused by similar flaws.

Many other disorders, though seldom directly caused by genetic abnormalities, have an inherited component. Heart disease, schizophrenia, arthritis, and our risk of developing many kinds of cancer all become more likely, or more severe, if we carry one or more deviant genes. Even our susceptibility to the virus that causes AIDS seems to depend on the makeup of our DNA.

Ashanti DeSilva inherited a defective gene that deprived her of an enzyme called adenosine deaminase (ADA). The result is a condition known as severe combined immunodeficiency, or SCID. Like a sort of inborn AIDS, SCID leaves its victims without a functioning immune system. This is the so-called bubble-boy disease, nicknamed after the Texas youngster known only as David, who lived out his life in protective sterile rooms and sealed garments like child-size spacesuits. There are several varieties of SCID. David carried a different genetic flaw than Ashanti does, but its effect was the same. Lacking any protection against disease, SCID patients can die of infections that other people would scarcely notice. A common cold can kill a boy like David or a girl like Ashanti. David died in 1984, at the age of twelve, of an infection caused by a failed attempt to cure his illness. Unlike David's form of SCID, Ashanti's can be treated. The girl received frequent doses of ADA, her missing enzyme, and thus could survive without the stringent protection David required. Yet even with treatment, her immune system functioned poorly. She was always sickly, and she could look forward only to a circumscribed life and an early death.

The only possible cure was to give Ashanti's body its own source of adenosine deaminase—cells equipped with a normal gene for the enzyme. The cells Ashanti received were her own, but they had been genetically modified. Injected into her bloodstream, they produced the missing ADA and allowed her immune system to function normally. It took four infusions over as many months to complete the cure, and Ashanti has required occasional booster treatments since then. But the procedure worked. Today, for all practical purposes, she is a healthy little girl. Several other patients have received the ADA gene since then with equally good results.

That success triggered a small avalanche of attempts at gene therapy. Dr. Anderson estimates that to date several hundred people have been treated with artificial DNA in hope of curing inherited illness or enabling their body to fight off diseases influenced by their genes. By late 1995, researchers were testing possible genetic cures for cystic fibrosis, hemophilia, rheumatoid arthritis, familial hypercholesterolemia, AIDS, and no fewer than sixteen different types of cancer. Most of these trials deal with working versions of the natural genes found to be defective in hereditary disease. However, in attempting to cure AIDS, or at least to slow its progress, researchers have designed their own gene to interfere with the reproduction of HIV, the virus that causes AIDS. Many tests of gene therapy for malignancy use another artificial gene intended to make the tumor cells more sensitive to anticancer drugs. We expect to see many more such creative experiments in the years to come.

There is no real trick to making the DNA required to cure almost any inherited illness. As long as doctors know which bit of hereditary material is defective, they can make a healthy version using standard gene-splicing methods. Over the last fifteen years or so, the techniques have become downright routine.

The problem is to insert the new genes into the cells that need them. Fortunately, nature gave researchers a prototype delivery system to work with. Viruses are just bits of DNA (or, in less common cases, RNA) wrapped in sheaths of protein. They have no reproductive system of their own. In order to replicate their kind, they fasten themselves onto the surface of a host; inject their DNA into the cell, using their surface proteins as a kind of syringe; and commandeer the host cell's machinery to make more of their own hereditary material and proteins. The gene doctors have adapted this system for their own use: They put their synthetic DNA into a viral protein sheath and mix these customized viruses with the target cells; in Ashanti's case, the cells were so-called T-cell lymphocytes, a critical component of the immune system. When the "viruses" try to infect the cells, what they inject is the health-giving artificial gene. This is the system Dr. Anderson and his team used. To date, it is still the most common method of gene delivery and probably the most effective.

There are other possibilities, however. Some researchers are working with small rings of DNA known as plasmids. It turns out that cells absorb plasmids spontaneously. If therapeutic genes are inserted into the plasmid, they too are absorbed. Other investigators have used liposomes, hollow droplets of fat that cells also have proved willing to take in, to smuggle genes into their target cells. Liposomes especially have advantages, because they are much simpler to make than modified viruses. Both these techniques are still in the early stages of development, however. In another five years, we should have a better idea of how they can best be used.

By then, researchers should also have mastered a new method of gene therapy for diseases such as SCID. It will be a major improvement on today's primitive procedure.

To date, most of these experiments have worked with blood cells, just as Dr. Anderson did; after all, they are easy to collect and return to the patient's body. Even for blood diseases, however, this process has an unfortunate limitation. Blood cells do not survive long; the body discards literally billions of worn-out blood cells every day. As the altered cells wear out and are cleared from circulation, all those precious therapeutic genes disappear with them. This is why Ashanti DeSilva still requires periodic booster treatments.

The solution is not to work with blood cells themselves but with the more primitive cells that manufacture blood cells—the stem cells of the bone marrow. Give stem cells a functional gene for, say, ADA, and all the new blood cells they make will contain that enzyme for the rest of the patient's life. It took years for researchers to master the task of collecting stem cells efficiently, but that obstacle finally has been overcome. Now the problem is to get artificial genes into the stem cells and to induce the stem cells to resume their job of creating new blood cells. Neither step has proved easy.

In 1993, a team at Children's Hospital, in Los Angeles, scored the first success in this effort. Dr. Donald B. Kohn and his colleagues managed to insert functioning ADA genes into the defective stem cells of three newborn SCID patients. Two years later, the children all were healthy. At least one other group has since reported similar results.

The procedure worked specifically because the patients were infants. During early childhood, when people grow the fastest, their stem cells divide rapidly to provide blood for the expanding body. Later in life, the stem cells are less active. To date, no one has cured an adult patient of any disorder by treating the stem cells. That obstacle is not likely to last long, however. Soon after the year 2000, gene implants into the stem cells of adults should become almost routine.

To heal genetic diseases outside the bloodstream, doctors must work with tissues even less tractable than stem cells. For example, to heal cystic fibrosis, doctors must supply working DNA to respiratory cells, which cannot be removed for treatment. Until recently, it seemed that the obvious solution was to insert the necessary gene into an adenovirus, which naturally infects the lung and nasal passages. It worked in the lab. But when Dr. Michael R. Knowles and his colleagues at the University of North Carolina, Chapel Hill, tried it, the procedure failed. Less than 1 percent of the target cells actually received any of the new DNA. Adenoviruses simply are not very good at delivering their cargo.

At Ohio State University and at Britain's North East Wales Institute, other researchers ran into a similar problem when they tried to cure Duchenne muscular dystrophy. In this illness, a flawed bit of DNA robs its victims of a crucial muscle protein called dystrophin. The scientists inserted artificial dystrophin genes into muscle cells. Then they injected the treated cells into the arm muscles of twelve afflicted boys. After six monthly treatments, none of the boys had improved.

These failures were disappointing for scientists working in the field but not entirely unexpected. As one would-be gene therapist commented, "You've got to remember that this technology is still in its infancy. It will be a long time before we get it all figured out."

Early in 1996, an advisory panel convened by NIH director Harold E. Varmus put it even more strongly. Stuart H. Orkin, of Harvard University Medical School, and his colleagues on the committee declared that the entire field of gene therapy had been oversold by both scientists and journalists; practical gene therapy may be farther off than most laymen imagine. The primary obstacle, the group felt, was simply getting therapeutic genes into the tissues that need them. To date, viruses have not been nearly reliable enough at it to make for practical patient care.

Nonetheless, we expect to see this and other problems solved well before the twenty-first century completes its first decade. In the next ten or fifteen years, dozens—more likely hundreds—of advances in gene therapy will be announced. Many of these breakthroughs will be translated almost immediately into practical healing methods for diseases that today are difficult or impossible to treat effectively.

One powerful boost to gene therapy is the Human Genome Project, a fifteen-year, $3 billion attempt to identify all the genes it takes to grow a human being and to learn their precise chemical nature. In its first five years, this effort proved to be the most successful "big-science" program since the days of the Apollo space launches—this despite chronic penny-pinching by an American Congress unsympathetic to public funding of— well, just about anything. The program's first major target, construction of a medium-resolution map of the human genetic material, was reached two years early. And thanks to newly developed mapping techniques, scientists should know the location—if not the precise chemical nature— of almost every human gene by the time this book appears. It is exactly this kind of unanticipated technical advance that invariably emerges from—and justifies—giant research programs such as the Human Genome Project.

We have already seen the first practical payoff from this largely theoretical endeavor. Since the Human Genome Project got under way in 1990, the pace of genetic discovery has accelerated dramatically. In little more than two years between early 1993 and mid-1995, researchers announced that they had located the genes for Huntington's disease, amyotrophic lateral sclerosis (also known as Lou Gehrig's disease), adrenoleucodystrophy (the devastating illness that served as the basis for the movie *Lorenzo's Oil*), a common kind of colon cancer, an inherited form of breast cancer, one variety of hypertension, and a hereditary form of obesity that markedly raises the victim's risk of developing severe diabetes—and that is just a partial list. Discoveries like these will be announced almost weekly for years to come. Many of these theoretical breakthroughs will give rise to practical treatments as soon as researchers can find the time and funding to follow up on them.

It is much easier to recognize that kind of large-scale trend in science than to forecast specific discoveries. Still, certain developments now seem almost inevitable. In the next ten or fifteen years, the gene doctors will learn to cure adults with Ashanti DeSilva's form of SCID. They probably will not cure David's form, however; it seems to be a much more intractable problem. They will cure cystic fibrosis, in which the patient lacks a crucial enzyme. Coping with sickle-cell anemia will be much more difficult, because patients will retain a defective, symptom-causing form of hemoglobin—the blood's oxygen-carrying protein—even if a new gene allows them to produce a fully functional version as well. Research will yield a genetic treatment for AIDS, but it will not cure this fatal infection, merely slow its progress. And the number of patients treated with artificial genes will leap from a few hundred to several thousand.

Gene therapy is not the ultimate answer to human suffering. It has strengths and limitations, like any other medical technology. And some forms of gene therapy will fire ethical debates that may prevent their use, at least in the United States, where religious and ethical concerns are a potent force in forming social and medical policy. The day will come— sooner rather than later—when doctors can alter the genetic code of the human egg and sperm, so that babies need never be born with illnesses like SCID and cystic fibrosis. The possibility of genetic tampering for trivial reasons—to choose the color of a baby's eyes, for example, or of its skin—will make it very difficult to put this theoretical ability into practice. Yet the ability to fix even some of nature's mistakes will cause the most profound transformation in health care that humanity has ever experienced. Fifteen years from now, those changes will be more than forecasts. We will have begun to see them for ourselves.

THE END OF CANCER

In 1902, an embryologist named John Beard conceived the notion that a tumor resembled a fetus growing within its mother's womb. Recently, a seventy-year-old biochemist named Hernan Acevedo has proved that Beard was right in at least one crucial way. And by 2002 that discovery may reduce most cancers from a terrifying disease to an irritating nuisance.

Make no mistake. We are going out on a limb here, much farther than we ordinarily would think of doing. Yet one advisor whose judgment we have relied on for years is a respected cancer specialist at a major eastern university. And what he cannot say in public, for fear of damaging his reputation as a conservative scientist, he states freely in private. "This is the end of cancer," he tells us. "In ten years, any family doctor will be able to treat most cases in his office."

There is a lot of research still to be done before that forecast becomes reality. But thus far it looks like he is probably correct.

To date, doctors have had to treat cancers individually. That is, they give one combination of drugs for breast tumors, other combinations for different forms of lung cancer, still other drugs for prostate tumors. And some malignancies are best approached with surgery followed by radiation treatments. It is as if cancer were not one disease, but a large group of illnesses that just happen to resemble each other. In fact, that is how many cancer specialists view the tumors they treat—as a broad spectrum of disorders that vary from one organ to another, and often from one patient to the next.

Yet it is intuitively obvious that cancers must be closely related. Look at their behavior. All tumors, no matter the organ or patient, start as a tiny mass of cells; almost surely, they begin with a single cell that somehow goes wrong. That adult cell ordinarily multiplies slowly, as more of its kind are needed, or it does not multiply at all. Yet suddenly it loses its adult inhibitions and begins to multiply much as embryonic cells do. The tumor grows slowly at first; some continue to smoulder this way for years. But with most cancers, it seems that at some point a critical mass is reached. Growth suddenly accelerates, and the tumor begins to shed cells. Those malignant cells colonize other organs and spread cancer throughout the body. Almost all tumors follow this pattern. Surely they must have some factor in common.

According to Dr. Acevedo, who works at the Allegheny-Singer Research Institute, in Pittsburgh, that factor is a hormone called human chorionic gonadotropin, or hCG. This hormone also connects cancer with pregnancy.

Ordinarily, hCG appears only in the fetus and in pregnant women. Within ten days after an egg is fertilized, it blocks the mother's menstrual cycle and triggers the production of other hormones that ready the womb to sustain the growing embryo.

The hormone hCG has another function as well. For many years, scientists wondered how the fetus makes peace with its mother's immune system. After all, it contains the father's genes as well as her own. If the woman's immune defenses were doing their job, they would reject the developing baby, just as they would a transplanted organ. And, as it turns out, her immune system does try to destroy the embryo. It fails because the infant is coated with hCG. Like many molecules, hCG carries a powerful electric charge. So do the white blood cells used by the mother's immune system to eliminate foreign matter. Both charges are negative, so the fetal and immune cells repel each other, like the identical poles of two magnets. The immune attack fails because the white blood cells cannot get close enough to harm the fetus.

Doctors have known for decades that hCG appears in the blood and urine of some cancer patients. The number varied widely from one study to the next. Some researchers found it in only 15 percent of patients; others said it occurred in three out of four. And there was always an unanswered question: Was it the cancer that produced hCG, or did the patients' bodies make it in response to the tumor?

Dr. Acevedo has finally found the answers. After twenty years of work, much of it spent in perfecting more sensitive tests for hCG, he has found the hormone in every cancer he has examined—seventy-eight cell lines in all. It appears in cancers of the breast, cervix, prostate, bladder, colon, pharynx, bone, muscle, and two kinds of lung cancer. At this point, there was little doubt that the hCG found in cancer patients was part of the disease itself.

To make absolutely certain, Acevedo retested twenty-eight kinds of tumor cell. Instead of looking for hCG, this time he searched for RNA made from the gene from which the hormone is manufactured. That evidence of gene activity would be proof that the cancer cells were actually producing the hormone, not just absorbing it from the mother's body. Acevedo found the RNA in all twenty-eight samples. This time there could be no doubt. Like the fetus, cancer uses hCG to defend itself against the immune system. As Acevedo says, "The hormone that gives us life also kills us."

There are at least four ways to put this theoretical knowledge to practical use, and cancer specialists will perfect them all over the next ten years.

Suddenly, doctors have a convenient test for cancer. If they find hCG in blood or urine, the patients is either pregnant or suffering from cancer. And as most cancer patients are either men or postmenopausal women, and many of the remainder are children, it should be an easy diagnosis to make. We expect an hCG-based test for cancer to reach the U.S. market no later than 1999.

Those tests will be based on monoclonal antibodies—absolutely pure immune markers that react to hCG with exquisite sensitivity, but to absolutely nothing else. Acevedo has developed one; other researchers have them as well. And monoclonals have another use: Tag them with a radioactive isotope, and you can trace them through the body and find out where they congregate. So a radio-labeled hCG antibody will not only identify cancer, it will show what organs are affected. These tests will be so sensitive that they should highlight, say, a breast tumor before it can be detected by hand or by X ray. They should be available by 2001 or so.

With effective antibodies, it is a short step from diagnosis to treatment. At first, antibody-based cancer treatments will be used to backstop conventional therapy. Such "adjunctive" treatments are surely needed.

One major problem with cancer surgery is that doctors can never be quite sure they have removed all the tumor cells. And if even one cancerous cell is left in the patient's body, it will continue to multiply and the tumor eventually will come back. This is why many cancer patients die of recurring tumors, even after what seemed to be a successful operation.

Soon, after cancer surgery, doctors will be able to give their patients anti-hCG antibodies. These monoclonals will clear away the hCG, so that the patient's immune system can attack and destroy any remaining tumor cells. That should almost completely rule out any recurrence of the disease. This treatment too will be available in the first few years of the next decade.

There is another hormone linked to cancer, a material known as epidermal growth factor (EGF). Normally, it promotes the multiplication of skin cells. In tumors, it promotes the multiplication of cancer cells. More than twenty years ago, doctors tried giving patients antibodies against EGF to make chemotherapy more effective. It worked to some degree, but there were problems. EGF is not nearly as abundant in cancers as hCG, so antibodies against it are not as effective as anti-hCG should be. And back then doctors had only animal antibodies to work with. The first time you inject, say, mouse antibodies into a patient, they attack the tumor. The second time, the patient's immune system attacks them and destroys the antibodies before they can take effect. Today, doctors have antibodies grown from human cells. These "humanized" antibodies are much more effective than animal-based agents and much less likely to cause rejection.

So there is the fourth use: Combine modern anti-hCG antibodies with chemotherapy, just as doctors once attempted to use anti-EGF. By all current evidence, it should make chemotherapy dramatically more effective. This technique too should reach the clinic five or six years from now.

That leaves one more possibility, in many ways the most promising use of all. Why not immunize patients against hCG, just as children are immunized against polio? Then their bodies would churn out their own native antibodies, and any tumor cells would be under constant attack. With a little luck, this could be a one-shot cure for cancer, available in any doctor's office. Tests are already under way.

Dr. Warren Stevens, of Ohio State University, already has a vaccine against hCG. He developed it for use as a contraceptive. Give a woman this vaccine, and any hCG is destroyed before it can prepare the womb to carry a fetus. Reportedly, the vaccine blocks pregnancy for about six months. In animal studies, it also helps to fight off several forms of cancer.

To date, there has only been one test of the hCG vaccine in human patients. It was a brief study, intended only to make sure the vaccine was safe to use. However, Dr. Pierre Triozzi, also at Ohio State, con-firmed that just one dose caused cancer patients to manufacture anti-bodies against hCG, without harmful side effects. And though patients were in the advanced stages of cancer and there was no real hope of curing their disease, Dr. Triozzi and his colleagues did note that the patients' tumors shrank noticeably after a single treatment. Continuing therapy should be much more effective.

More extensive tests, using the Stevens vaccine against cancers of the colon or pancreas, are just getting started. Within a year or two, we should know for certain just how effective treatments based on hCG really can be.

Again, we expect this therapy to be available within the next five years or so. By 2010, it should be standard practice. Very probably, a diagnosis of cancer will be much less terrifying than it is today. Just possibly, cancer will be no more than an annoyance—a once-deadly disease that, like tuberculosis, medicine has learned to cure.

If we sound unusually optimistic about this, it is only because we have caught the enthusiasm of people much more knowedgeable in this field than we will ever be. One of the brightest, most versatile cancer specialists we know is Dr. William Regelson, of the Medical College of Virginia. When Acevedo's most recent work was published in *Cancer*, it was Dr. Regelson who wrote the prestigious journal's editorial to alert other doctors to the importance of the work.

"I'm seventy years old," he comments. "I have seen so many 'miracle cures' come and go that it is hard to get excited about these things. But this is an incredibly dramatic development. The cancer-research community ought to focus its attention on this. As far as I am concerned, if Acevedo's work holds up, it's all over but the shouting."

THE FOUNTAIN OF MIDDLE AGE

If you have been secretly hoping to stay young forever, science still cannot help you. But if you can bear to settle for a bit less, then the answer just might be at hand. It does seem likely that we can now enjoy extra years, and perhaps even decades, of health, energy, and an appearence that would fire the vanity of an average person in his forties or fifties.

Doctors have known for many years that our supply of crucial hormones falls, sometimes dramatically, as we get older. Now they are discovering that the loss of hormones is not just a symptom of aging; it actually causes many of the changes that we know as growing old. In some cases, at least, hormone supplements can slow or reverse those symptoms. There even is tantalizing evidence that one formerly obscure hormone can slow aging itself and perhaps lengthen our lives by as much as 30 percent.

This third revolution will touch all our lives. By the age of forty, almost everyone suffers from at least one chronic disease—arthritis, diabetes, heart disease, perhaps a hidden prostate cancer—brought on largely by our advancing years. And that is just the beginning. If nature has its way, most of us will spend the last years of our lives in a nursing home, the last weeks in a hospital bed strapped to the costly, uncomfortable, and ultimately futile machinery of high-tech life support. In the United States alone, more than 7 million people already require long-term care. If hormone replacement offers to free us from the ills of old age, as it certainly seems to, then this may be the biggest medical revolution of all.

The best-known version of hormone replacement therapy, of course, is the combination of estrogen and progesterone given to women during and after menopause. Originally meant only to combat the "hot flashes" and irritability of menopause itself, estrogens have proved to retard or eliminate many disorders that often appear as women grow older. These include vaginal dryness, weight loss, and drying and aging of the skin. Far more important, postmenopausal women who take estrogens are less likely to develop osteoporosis and much less likely to suffer from heart disease. Though some doctors, and many of their patients, fear that estrogens may have a dark side, years of careful researcher still have not given us clear proof that estrogens significantly raise a woman's risk of developing breast or uterine cancer. The danger, if it exists, must be much smaller than the pessimists believed. Certainly it seems to be outweighed by other findings: According to recent studies, estrogens cut in half the death rate of postmenopausal women at all ages and from all causes. Next to vaccines, that makes estrogen replacement one of the most powerful disease-preventing medical procedures ever discovered.

Of course, the dramatic success of estrogen does not mean that taking hormones always will prove equally beneficial. Yet it seems a powerful argument for at least testing other forms of replacement therapy. And in recent years, medical scientists have taken the hint. One program run by the U.S. National Institute on Aging (NIA) has nine research teams studying the possible benefits of growth hormone (GH) in our middle years and old age. Other hormones being tested include GHRH, or growth hormone releasing hormone; testosterone; DHEA (dehydro-epiandrosterone), a product of the adrenal glands; and melatonin, from the pineal gland, which lies deep within the brain. They all seem to offer at least some benefits.

To date, growth hormone and its releasing factor probably have received the most clinical study. In childhood, growth hormone builds our bodies to their adult height and weight. And until recently, scientists believed that was its only function. The fact that our supply of growth hormone begins to decline in our thirties and forties seemed unimportant. The only medical use of GH was to help children grow to normal height when their pituitary glands did not naturally produce enough.

However, in the last five years, research has shown that GH continues to wield a powerful influence on our bodies even after we have stopped growing. As we age, we tend to lose our muscle tone; eventually our muscles become thin and stringy. Men especially begin to put on weight. Our skin becomes thinner and less supple. These changes occur even if we continue to exercise regularly, rather than spending our lives mesmerized by the television. Though other factors also may be involved, all these effects have been traced, at least in part, to the loss of growth hormone.

Replacement therapy can reverse them. Not long ago, Dr. Daniel Rudman and his colleagues at the Medical College of Wisconsin tested it in men between the ages of sixty-one and eighty-one. For six months, Dr. Rudman's patients received growth hormone three times each week. By the end of the test, all twelve had regained lost muscle, developed thicker and healthier-looking skin, and shed much of the fat that time had deposited around their waist. Several other tests have confirmed the Milwaukee group's results.

Growth hormone releasing hormone does just what its name implies. It is a chemical in the brain that triggers the pituitary gland to release growth hormone. Predictably, it also helps to reverse the symptoms of age. Dr. Robert S. Schwartz and his coworkers at the University of Washington, in Seattle, mirrored Dr. Rudman's results. In aging men, they found that GHRH markedly reduced body fat and stimulated muscle growth even when their patients failed to exercise. Best of all, there were no undesirable side effects at all.

About testosterone in aging, matters are less clear. However, this hormone is another obvious candidate for replacement therapy. There is no evidence that men experience the fabled "male menopause," in which their supply of testosterone suddenly plummets, as women's estrogen levels do. However, they do lose the male hormone as they grow older. By age eighty, half to two-thirds of their testosterone has vanished.

This probably is important. In several small studies, researchers have found that in men over age fifty, testosterone supplements improve muscle mass and strength, lower blood cholesterol, reduce angina, and may even cut the risk of heart attack. One of the most convincing reports comes from Dr. Frederick M. Ellyin, of the Chicago Medical School. In this work, ten healthy men between the ages of sixty and seventy-five received small doses of testosterone once every week or two for two years. At the end of that time, most had lost body fat, cut their cholesterol levels, and gained muscle strength. The men also reported more sexual interest and activity and a generally improved sense of well-being. A much larger study of testosterone replacement is now under way at the University of Pennsylvania School of Medicine, where doctors are giving testosterone to fifty healthy men over age sixty-five. Eventually, the men will be tested for improvements in such factors as bone calcium, muscle mass, and strength. Conclusive results will not be available until late in 1998.

One of the most interesting antiaging hormones is DHEA. Until the early 1970s, medical researchers knew little about this hormone, which is produced in the adrenal glands. At best, they thought it served as a raw material for the manufacture of some other, more important steroid. And, in fact, it turns out that most of the steroid hormones are derived from DHEA. But there was tantalizing evidence that it must have some greater significance. At the age of twenty-five, our bodies are flooded with DHEA; it is by far our most abundant steroid hormone. But DHEA levels decline steadily over time. By age seventy, we have only 10 or 15 percent as much DHEA as we did in our prime. Surely that must have some effect.

It does. The breakthrough came when Dr. Arthur Schwartz, of Temple University's Fels Institute, came across a little-noticed research report. Women who produce less DHEA than normal, it said, have a much higher risk of breast cancer. Dr. Schwartz already knew about another effect of DHEA: it prevents obesity in a strain of mice that, thanks to a genetic defect, naturally grows enormously fat. And that brought up an intriguing possibility. Since the 1920s, researchers in the field have known that they could double the lives of test animals—and incidentally cut their cancer rates—by putting them on a diet with plenty of vitamins and minerals but near-starvation calorie levels. It seemed to Schwartz that DHEA might work the same way. He has been following up on that idea ever since. In a long series of studies, he has never been able to lengthen the life span of test animals using DHEA alone. But the hormone does preserve many of the characteristics of youth. Mice receiving DHEA remain slender and active. They are glossier and markedly less gray than untreated littermates. And they are much less prone to develop cancer and the other diseases that usually strike both mice and men as we grow old. For almost twenty years, it has seemed likely that DHEA would dramatically improve human health in later life, just as it does that of animals.

Now other scientists have carried DHEA studies into the clinic. The current leader in this field is Dr. Samuel S. C. Yen, an endocrinologist at the University of California at San Diego. In one study, Dr. Yen gave DHEA to sixteen men and women in middle and old age. After one year of treatment, the men had larger, stronger muscles, denser bone, and less body fat. Women experienced no similar improvements. However, both men and women reported that they slept better and found it easier to cope with stress. Other human trials have shown that DHEA supplements improve failing memories, boost the immune system, and help to relieve the symptoms of autoimmune diseases such as lupus erythematosus. Clearly, there is much more to learn about the benefits of DHEA. Yet already it seems clear that this once-neglected hormone can powerfully improve our health in later life.

The newest candidate for replacement therapy is at once the most promising, the most popular, and the least well tested in formal human trials. This is melatonin. Following a wave of publicity in 1995, health-conscious readers have been downing this latest wonder drug so quickly that health-food and nutrition stores have had a hard time keeping it in stock. No wonder. Several innovative and highly reputable scientists have touted it as a cure-all for sleep disorders, jet lag, and stress; a shield against cancer; a tonic for the immune system. Studies even have shown that melatonin supplements dramatically extend the life span of animals. If it works for people, this will change the most fundamental aspect of human existence.

Until recently, melatonin was known only as a sleep inducer. When night falls, our pineal glands begin to produce the hormone, and we fall asleep. Many people who have trouble sleeping find that a small dose of melatonin at bedtime helps them drop off quickly and remain asleep all night. Some, perhaps one in three, also report that they need markedly less sleep, just a few hours a night. In fact, this hormone can do much more for us.

Melatonin has proved to be the most influential hormone our bodies produce. Throughout our lives, our melatonin levels follow a daily rhythm. They rise at night, peaking at about 2:30 A.M., and decline during the daylight hours. This cycle is the timing signal that keeps all our other bodily rhythms in train with the world around us. Our levels of growth hormone, thyroid, and many other bodily chemicals all vary throughout the day in response to melatonin. Flying across time zones causes jet lag because all these cycles suddenly are out of step with the day around us. They catch up at different rates, and we feel miserable until they all are synchronized again. Melatonin can speed that process. One dose at bedtime in our new time zone eliminates jet lag almost completely.

There are longer cycles as well. As the seasons change, our nightly melatonin peak varies slightly. Some researchers believe this longer-term change helps our bodies to adapt to the annual weather cycle. (In equatorial regions, where temperatures vary little, melatonin may be governed by, and help the body prepare for, the annual cycle of rain and drought.) Melatonin levels also vary with our time of life. Early in childhood, the pineal gland floods the body with melatonin each night. Shortly before puberty, our melatonin levels at the nightly peak drop abruptly. This timing is more than coincidence. The decline of melatonin actually triggers sexual maturation. As the years pass, our melatonin levels continue to decline. At age forty-five, we produce half as much at the nightly maximum as we did when we were children. By age eighty, scarcely a trickle is left, and the regular day/night cycle has become disordered. It would be strange indeed if the loss of so crucial a hormone did not have some profound effect on us.

In fact, according to a scientist named Walter Pierpaoli, it may be the cause of aging. Dr. Pierpaoli, an endocrinologist and immunologist, is research director at the Biancalana-Masera Foundation for the Aged, in Ancona, Italy. He has performed three key experiments that firmly link melatonin to the process that causes us to grow old.

In the first, he simply gave his aging mice a little extra melatonin at night, much as readers of his book are dosing themselves today. His test animals lived longer and retained the health and appearance of much younger animals.

Then Pierpaoli asked a colleague—an extremely skilled surgeon—to extract the pineal glands from young mice and implant them into old ones. Again, the treated mice lived longer and remained physically "younger" than their littermates.

Most recently, he has transposed the pineal glands of young mice into old animals and old pineal glands into young mice. Almost immediately, the young animals began to look older. Their fur became gray and patchy, their eyes developed cataracts, and they took on the stiff, wobbly gait of much older mice. They died of what looked like natural old age well ahead of schedule. By comparison, elderly mice with young pineal glands, producing youthful quantities of melatonin, regained their energy; their gray, balding coats grew shiny and thick again; and they scampered—and reproduced—like mice in their prime. And the chronologically old mice lived longer. As Pierpaoli and his colleagues have gained practice at the transplant procedure, the implants have stretched the lives of mice by up to 40 percent.

Melatonin alone is not quite that effective; it seems the pineal gland must produce other factors that improve health and extend life. Yet the hormone gave his treated mice the equivalent of twenty extra years of life, in human terms. And they kept their youthful health, energy, and appearance until the end of their lives.

Pierpaoli emphasizes that melatonin does not work its miracles as a drug might, by tampering with animals' biochemistry. Not everyone agrees, but the evidence is on his side. One opposing theory holds that we grow old because chemicals known as free radicals attack and destroy our proteins, DNA, and other material crucial to life. According to this idea, almost any nontoxic substance that eliminates free radicals might preserve our health and extend our lives. Recently a cell biologist named Russel Reiter, who has devoted a long and productive career at the University of Texas to the study of melatonin, discovered that this hormone is the most powerful destroyer of free radicals yet discovered. This, he believes, explains melatonin's dramatic benefits. However, Pierpaoli ruled out that possibility even before Dr. Reiter made his discovery. In one experiment, he divided his test mice into two groups. Some received their melatonin at night, when their levels would naturally be at their highest. The remainder took it during the day, when light, the digestion of food, and other factors create the most free radicals. If melatonin functioned only as a free-radical inhibitor, the mice receiving it during the day should have had the greatest benefit. Instead, they aged and died on schedule. The animals receiving melatonin at night remained young and lived longer. The body's natural timing pulse, that melatonin peak in the hours after midnight, proved all-important.

Pierpaoli's work deals with mice, not men. And though most scientists assume that human beings age as mice do, there is no guarantee. There could be some as-yet unknown complication that prevents melatonin from extending our lives. Only human experience will answer all the questions that remain about this promising new hormone replacement, and it will take many years to accumulate.

Yet there is good reason to believe that melatonin offers at least some valuable benefits. As we age, our immune systems lose their ability to protect us against disease. One reason is the failure of the thymus, a small gland tucked away at the base of the throat. Its principal hormone is thymosin, which plays several crucial roles in our immune defenses. As we age, the thymus shrivels to perhaps one-fourth of its original size, and thymosin almost disappears—unless we receive melatonin supplements. In animals, at least, melatonin preserves the thymus and keeps the immune system strong.

There is evidence that it works in people as well. Several years ago, Dr. Paolo Lissoni and his colleagues at San Gerardo Hospital, in Monza, Italy, tried supplementing an experimental cancer drug with melatonin. The drug was interleukin-2 (IL-2), a natural hormone from the immune system that is involved in one of the defense mechanisms that also depends on melatonin. Ordinarily, IL-2 is effective only at extremely high doses—so high that patients suffer from high fevers and severe nausea. Combined with melatonin, the drug was much more effective against tumors, and it worked at much lower dosages. It appears that melatonin had restored a crucial part of the patients' immune systems. This seems a powerful hint that melatonin could also shore up the failing immune defenses of healthy seniors.

As research in hormone replacement continues, science is likely to discover other therapies that could help keep us healthy and well. The latest candidate for supplements is leptin, a hormone secreted by fat cells that shuts off appetite when we have eaten enough. It was discovered late in 1994 by Jeffrey M. Freidman and his fellow researchers at the Howard Hughes Medical Institute of the Rockefeller University. In experiments, leptin supplements have caused dramatic weight loss in several groups of obese mice. Some were overweight because of a genetic defect that stops their cells from making functional leptin. Others had been fed a high-fat diet. The rest were normal mice. In just four days, even the normal mice lost nearly all of their body fat, about 12 percent of their total weight. If human beings respond as well, obesity could soon be a thing of the past. It may be, however, that overweight people still produce enough leptin; their brains just no longer respond to it. In that case, researchers may still be able to devise a drug to restore the body's lost control mechanism. We will know the answer well before the turn of the century.

There are many questions to be answered, and practical problems to be solved, before we gain all the benefits that hormone replacement has to offer. We are just beginning to understand how to use DHEA and melatonin, and even less is known about leptin. Excessive dosages of growth hormone can aggravate diabetes and some other disorders. And it costs about $14,000 per year to provide growth hormone for just one elderly patient. Yet research can overcome this kind of obstacle, and that research is being done. Fifteen years from now, we will have learned how best to use these and other hormones. And thanks to replacement therapy, we will escape most of the diseases that have plagued the elderly of past generations. We may not even grow old.

PIG HEARTS AND MONKEY MARROW

For years, medicine has ground out slow, steady progress in organ transplantation. Since the first, seemingly miraculous heart transplant performed by Dr. Christiaan Barnard in 1967, the field has grown to the extent that several thousand patients receive the hearts of human donors each year in hospitals around the world. Whole clinics are devoted to kidney transplantation, which has become the treatment of choice for many patients. A few patients even have received animal organs to tide them over. Not long ago, the newspapers were full of slightly breathless accounts of an AIDS patient who received bone marrow from an ape in an effort, ultimately unsuccessful, to preserve his failing defense against infection. To date, almost all transplants have required harsh immune-suppressing drugs to prevent the patient's body from rejecting its new organ; the AIDS patient, whose immune system had already been almost destroyed, was the only exception we know. This is a profound disadvantage, because it leaves the patient with a kind of artificial AIDS, in which even a simple cold can turn life-threatening. And it seems that immune suppression often fails in the end. After months or years, the donated organ is lost, and the patient dies.

That is about to change. One rather remarkable scientist has discovered that much of what we thought we knew about the immune system was wrong, or at least woefully incomplete. He has developed a technique that allows doctors to transplant organs from one person to another without all the cumbersome tissue typing now used to ensure that donor and recipient are "compatible." Doctors using his method can even transplant organs from a rat into a rabbit, or a monkey into a man, without fear of rejection.

Unfortunately, we can offer few details of this research. The scientist pioneering this technique, whom we have known for several years, has asked that we keep even his name confidential until other scientists have had the opportunity to confirm his work. So far, he has tested his work only with transplants of bone marrow, mostly in rodents. However, the technique should work equally well with other organs and tissues. Independent experiments with bone marrow in dogs reportedly began late in 1995.

This research still has a long way to go before the first human trials, and it will be even longer before human cardiac patients routinely receive the hearts of pigs, if it proves successful at all. Yet medical research progresses so quickly that we suspect it will reach clinical practice no later than 2010.

As of May 5, 1993, the most recent date for which we can find comprehensive figures, no fewer than 31,279 people in the United States were waiting for compatible donor organs—23,533 for kidneys, 2,843 for hearts, 2,654 for livers, 1,071 for lungs, 140 for pancreases, and the remainder for both a kidney and a pancreas or a heart and lungs. By May 5, 2010, transplant patients will wait just long enough to schedule a surgeon and an operating room.

THE STRANGE STORY OF OMENTAL TRANSPOSITION

Our fifth development, not quite a revolution, involves a treatment that human patients have undergone for nearly two decades. It has taken that long to win a fair hearing for this remarkably versatile procedure, but it begins to appear that omental transposition will finally take off in the next ten years. If so, we suspect that many patients will be the better for it.

There is nothing terribly impressive about the omentum. It certainly does not seem like a subject for controversy. It is just an extra little flap of the tissue that lines the abdomen and wraps better-known organs like the stomach and intestines. Yet it may be one of the most valuable single discoveries in modern medicine. It has also earned one of its pioneers an exposé on the television newsmagazine "20/20"—"a hatchet job," as one of the man's friends bitterly describes it.

For more than twenty years, surgeon Harry Goldsmith has been moving the omentum from place to place within the body. He dissects it away from the abdominal wall and clips it here and there so that it can be stretched, rather like a string of paper dolls. This allows him to shift it as far away as the skull, without removing its connection to the arteries and veins that supply its blood and remove the fluids it absorbs. This can be an easy operation or a very difficult one, depending on where the omentum is to end up.

Dr. Goldsmith describes omental transposition simply as "a new delivery system for putting things into various parts of the body." The omentum is richly supplied with blood, neurotransmitters, and nerve growth factor, a hormone whose function the name captures well—it causes nerves to grow wherever it appears. These are the materials that the omentum delivers wherever it is placed, and the body often finds them very useful. The results can be too astonishing for some observers to accept.

"I did a boy in Germany several months ago," Goldsmith says, with a quiet humility that would have shielded a less creative doctor from attack.

When he was two years old, he had drowned in a pond next to his house. The rescuers had kept him alive, but four years later he remained in a vegetative state. His parents had heard that some of these patients improve when placed in a hyperbaric chamber, which forces extra oxygen to the brain. Not all of their brain nerves are dead. Some are barely alive, but they begin to function again when they receive enough oxygen. So they took him for hyperbaric treatments, and he seemed to improve slightly while in the chamber.

I watched him for a year or two and warned the parents that operating probably would not do much good. Eventually, they decided it was the only hope they had, and I went ahead with it. Since then, his mother has called me almost every other day to report improvements. He clearly is not a normal six-year-old, but he is awake and active. Two months after the operation, he is feeding himself with a utensil.

That kind of miracle story deserves skepticism, but Goldsmith is not the only one telling such stories. At the First International Congress of Omentum in the Central Nervous System, held in Xuzhou, China, in March 1995, no fewer than 162 research teams submitted papers for presentation. Sixty-two were selected for publication. Many of those reports were almost as remarkable as the story of Goldsmith's young drowning victim.

Professor Sigfried Vogel, of Charite Hospital, in Berlin, told of placing the omentum directly onto the brain of a man with intractable Parkinson's disease. Parkinson's patients develop uncontrollable tremors because a critical part of the brain loses its ability to produce a neurotransmitter called dopamine. The omentum is well supplied with neurotransmitters, including dopamine. After surgery, Vogel used a PET scan to measure dopamine levels in the afflicted part of the patient's brain. They were significantly higher than before, and the patient's trembling was much reduced. Only one other treatment is open to Parkinson's patients who do not respond to drugs: an implant of fetal nerve tissue. Omental transposition, if it works equally well in future patients, seems a better alternative.

One of Vogel's colleagues at Charite Hospital, a Dr. May, reported on a similar operation in a patient with severe epilepsy that had resisted drug therapy. Before the operation, the patient was struck by as many as twelve seizures every hour. Two months later, the rate was down to one every other day.

"We know that brain nerves become electrically unstable if they do not receive enough oxygen," Dr. Goldsmith speculates. "It may be that the omentum is providing enough extra oxygen to let them function normally." However, he is quick to point out that an apparent success with one patient is meaningless; it will take extensive research before doctors can conclude that omental transposition really helps patients with epilepsy or with Parkinson's disease.

Dr. Larry Ferguson is chief of neurosurgery at Chicago's Michael Reese Hospital. He presented a paper on arachnoiditis, a condition in which scar tissue from a previous back injury presses on the spinal cord, causing horrific pain. Removing the scar tissue eases the pain only temporarily; patients soon relapse. However, in six patients, Dr. Ferguson first removed the offending tissue and then placed the omentum over the wound. In three patients, the procedure seems to have banished the problem permanently. Two got slightly lesser benefits, and one was rated only "fair" but had at least some improvement. Three more patients also showed marked improvements, two with persistent leaks of the fluid that bathes the spinal cord and brain, and one with recurrent cysts that pressed on the spinal nerves.

Several other papers dealt with spinal injuries. As far as most neurologists are concerned, the spinal nerves cannot repair themselves after injury. Yet Dr. Thomas Yarrow, a veterinarian from London, told of dogs that had been struck by cars, their backs broken. Ordinarily, when the animal has no sensation in the hind legs, veterinarians assume that it is beyond help. But when Dr. Yarrow transferred the omentum to the injured area of the spinal cord, five of nine such dogs walked within three weeks. Another could walk again within a year. Could omental transposition help some human paraplegics to walk again? It seems possible—and worth finding out.

In other countries, research on omental transposition is growing rapidly. Yet in the United States, where Harry Goldsmith almost single-handedly has rescued an old, half-forgotten operation from obscurity, it remains almost unknown—save for the occasional exposé.

As for Goldsmith himself, he is in semiretirement. After years at such prestigious institutions as Boston University and the Sloan-Kettering Cancer Center, he appears to have given up on finding an American research center willing to give his work the support it needs. Instead, he travels often to other countries—Germany, Italy, Spain, China—anywhere a patient needs help and an interesting case might provide more information.

As Dr. Goldsmith points out, it is much too soon to draw any final conclusions about omental transposition. Yet research on the controversial procedure is continuing, if not in the United States where it originated, at least in other lands. In fifteen years, we should have a much better idea of its value. We will not be surprised if it proves to be a bright new hope for patients with spinal injuries and many other disorders that cannot be treated today.

APPENDIX:
A TIMETABLE FOR
THE FUTURE

Each year, Professor William E. Halal and his students at George Washington University perform a Delphi survey that attempts to forecast the development of emerging technologies. This appendix reports the results of his 1996 poll.

For those who are not familiar with the Delphi method, it is one of the most popular techniques in scientific forecasting. It consists of two successive polls about the same subject. First, experts in the field are asked to give their expectations about the topic under study. These preliminary answers are reported back to the panel members. Then, having considered the views of their colleagues, the experts are allowed to revise their answers. In this way, a consensus forecast is developed. The results of this second stage usually agree much more closely than the first responses did. In the long run, they also are much more likely to prove right. Over the years, many thousands of Delphi studies have been performed in fields ranging from politics to particle physics. Their record for correctly anticipating future events is unmatched by the results of other forecasting methods.

In the 1996 survey, Dr. Halal examined twelve fields: energy; the environment; farming and food; four segments of information technology—computer hardware, computer software, communications, and information services; manufacturing and robotics; materials; medicine; space; and transportation. In each area, he offered a brief introduction to the topic, suggested from five to nine possible developments, and asked the panelists four questions about them: How likely was each to occur? If it did come to pass, when could it be expected? What would it be worth in the global marketplace? And which nations are likely to lead in that field? The panelists were also invited to offer any comments they might have about each scenario.

What follows are the results of Dr. Halal's most recent survey. Table I contains Forecasting International's answers to the first phase of the Delphi poll. Table II presents the consensus view, which evolved from the second phase. Table III offers a timeline of future developments in these dozen crucial technologies. Table IV identifies the experts who contributed to the poll.

Table I omits our estimates of future market demands in the areas covered by this study. However, to give you a sense of the potential economic importance of the possible developments that follow, here are the sizes of the U.S. market in some existing industries, as of 1993:

Fiber optics	$2 billion
Supercomputers	$3 billion
Mobile communications equipment	$4 billion
Consumer electronics	$12 billion
Data networking equipment	$22 billion
Communications	$42 billion
Personal computers	$60 billion
Aerospace industry	$124 billion
Cable entertainment	$163 billion

In Table II, we have included the consensus estimates of future markets, each pegged to the date at which the innovation is expected to reach general use.

Here, then, are the results of Dr. Halal's 1996 Delphi Survey of Emerging Technologies. In Table I, the "confidence levels" attached to each field represent our degree of certainty that our predictions in that area will prove correct; Table II presents the average of the levels reported by the individual respondents for each specific event they were asked to evaluate. The occasional "authors' notes" in Table I have been added to account for things we have learned since responding to the survey. Some parenthetical additions have been made to certain scenarios to explain how we interpreted questions that might otherwise seem ambiguous or overly telegraphic.

In general, you will find that Forecasting International's predictions are on the conservative side. In nearly all scenarios, we project that the event has less probability of coming to pass or will be delayed longer than the consensus view predicts. We believe that this reflects our own background. Many years of dealing with business and political affairs have taught us that things generally take a lot longer than mere technical possibility would suggest, while some apparently reasonable concepts disappear entirely when forced to compete in the marketplace. In those rare cases where our forecasts are more optimistic than the consensus view, it usually is because we can see some strong competitive advantage that will accrue to the company or nation that develops the technology and thus believe that a strong research and development effort will be made in that field.

Dr. Halal's survey pushes the limits of what is possible in scientific forecasting. In most fields, we are much more comfortable with shorter-term predictions, in the range of ten to twenty years. The shorter the period, the easier it is to extrapolate the many influences that will feed into their ultimate outcome. In this survey, most scenarios deal with events that cannot be anticipated for three to five decades. Nonetheless, we believe that the forecasts below represent as accurate a preview of coming events as can be made at this time.

TABLE 1

FORECASTING INTERNATIONAL'S RESPONSE

I. Energy   Confidence level: 75 percent

Alternative sources for fossil fuels are now becoming feasible, as the cost of conventional energy and its resulting environmental damage grows. Photovoltaic cells are now commonly used for special purposes, such as calculators, billboard lights, battery chargers, and experimental solar-powered automobiles, and they are starting to be used for homes and power plants. Luz International Company operates a state-of-the-art solar power plant in the Mojave desert that produces electricity at comparable costs to conventional sources. Thousands of windmills are now producing energy in the United States. More efficient lights are available that cut energy use by more than 50 percent. More efficient fuel cells using gasoline, natural gas, and hydrogen are being developed for commercial use. According to the U.S. Department of Energy, renewable energy supplies about 10 percent of the United States' energy, and this figure should increase to 28 percent by 2030. Fission nuclear power is undergoing an important transition, as reactor designs become fail-safe and small enough to power large buildings.

Events:

1. *A significant portion (10 percent) of energy usage is derived from alternative energy sources, such as geothermal, hydroelectric, solar/photovoltaic.*
Probability: 50 percent
Year: 2025
Leading nation: United States
Comments: Fusion is more likely by 2030. Once it becomes available, none of the other alternative energy sources will be economically viable.

2. *Energy efficiency improves by 50 percent through innovations in transportation, industrial processing, and environmental control.*

Probability: 30 percent by 2025
                        50 percent by 2040
Leading nations: United States, Japan

3. Fuel cells, converting fuels to electricity, are commonly used. (They provide 30 percent of the electric power generated.)
Probability: 50 percent
Year: 2025
Leading nations: United States, Israel, Japan

4. Biological materials, such as crops, trees, and other forms of organic matter, are used as significant (10 percent) energy sources.
Probability: 20 percent
Year: 2025
Leading nation: United States

5. Fission nuclear power is used for at least 50 percent of electricity generation.
Probability: 25 percent
Year: 2025
Leading nation: United States

6. Fusion nuclear power is used commercially for electricity production.
Probability: 50 percent
Year: 2030
Leading nations: United States, Russia

7. Hydrogen becomes routinely used in energy systems.
Probability: 50 percent
Year: 2050
Leading nation: United States

II. Environment   Confidence level: 40 percent

Global warming, acid rain, ozone depletion, and other such threats have caused government officials, business leaders, and households to rate the environment as the biggest issue facing this decade. (Authors' note: This is distinctly a minority position. In the United States, at the time this survey was begun, crime was generally rated the most important issue facing the country, followed by race relations and the need to balance the federal budget.) Many corporations are voluntarily recovering manufacturing wastes to avoid pollution and to recapture valuable byproducts, and they are redesigning products and packaging to reduce environmental impacts. For instance, Dow Corning saved more than $500 million through its "Pollution Prevention Pays Program," and DuPont is planning to curb all environmental emissions from its products. Japan recycles 50 percent of all household wastes, and the concept is being adopted in most American cities. As a result, the proportion of steel, aluminum, glass, paper, and other products composed of recycled materials is growing. Industrial ecology focuses on recycling and reusing byproducts within and between industries.

Events:

1. *The majority of CFCs are replaced by materials that do not damage the ozone layer, such as HFCs.*
Probability: 30 percent
Year: 2025
Leading nations: Japan, United States

2. *One-half of the waste from households in developed countries is recycled.*
Probability: 50 percent
Year: 2050
Leading nations: Japan, United States

3. Most manufacturers adopt "green" methods that minimize environmental pollution.

Probability: 50 percent

Year: 2050

Leading nation: United States

4. The majority of manufactured goods use recycled materials.

Probability: 50 percent

Year: 2050

5. Improvements in fossil fuel energy efficiency and greater use of alternative energy sources reduce "greenhouse" gas emission by one-half from current volumes.

Probability: 50 percent

Year: 2050

Comment: This forecast assumes that fusion power reaches the marketplace and provides a cheap source of electricity. Without that economic incentive, so large a reduction in greenhouse emissions is much less likely.

6. The majority of manufacturing facilities use industrial ecology (eco-industrial parks operating as a closed system) to reduce waste and pollution.

Probability: 30 percent

Year: 2025

III. Farming and Food   Confidence level: 75 percent

Hydroponic farming is increasing crop yields and can be used anywhere. Food is being manufactured from raw materials to produce artificial foods, such as the use of cellulose from wood to make bread, meats, etc. Aquaculture now provides about 22 percent of fish consumption, and is showing rapid growth as the demand for seafood continues to rise because of trends in health consciousness and medical research. The enormous potential for aquaculture is suggested by the fact that in 1991 livestock production was a $70 billion market, while aquaculture only accounted for $1 billion. In 1995, aquaculture is a $26 billion industry. New food derivations from sea vegetation are being developed. Advances in biotechnology have produced the first commercially available, genetically altered vegetables that grow faster, resist disease, and are more robust. Alternative/organic farming methods, used to rejuvenate soil, control crop pests, and recycle organic materials, are being adopted by many farms, especially as the use of fertilizers and traditional farming methods becomes increasingly expensive and less effective.

Events:

1. Genetic engineering techniques are routinely used to produce new strains of plants and animals.
Probability: 70 percent
Year: 2000
Leading nation: United States
Comment: Depending on how we choose to define the word "routinely," it could be argued that this already has come to pass.

2. The use of chemical fertilizers and pesticides declines by half.
Probability: 40 percent
Year: 2025
Leading nations: United States, Canada

3. The majority of farming in industrialized countries incorporates alternative/organic farming techniques into traditional methods.
Probability: 35 percent
Year: 2025
Leading nations: United States, Canada

4. Seafood grown using aquaculture provides the majority of seafood consumed.
Probability: 50 percent
Year: 2015
Leading nations: United States, Japan

5. Automation of farming methods, using technology such as robotics, is common. Thirty percent [of farming] employs some such technology.
Probability: 50 percent
Year: 2025
Leading nation: United States

6. Precision farming (computerized control of irrigation, seeding, fertilizer, pesticides) is common. Thirty percent [of farming] employs some such technology.
Probability: 60 percent
Year: 2025
Leading nation: United States

7. Urban production of fruits and vegetables using greenhouses and/or other intensive production systems is common. Thirty percent or more [of fruits and vegetables] are grown in such "farms."
Probability: 50 percent
Year: 2050
Leading nations: United States, Canada

8. Produce grown using hydroponic methods is common. (It accounts for 30 percent or more of all produce.)
Probability: 30 percent
Year: 2050
Leading nation: United States

9. *Artificial meats, vegetables, bread, and other foodstuffs are commonly consumed. (They account for 30 percent of all such products.)*
Probability: 25 percent
Year: 2025
Leading nation: United States

IV. Information Technology; Computer Hardware

Confidence level: 60 percent

There seems to be no end in sight to the 40-year trend in which computer power has increased about 50 percent per year, or by a factor of ten with each successive generation of machines. Microcircuitry now commonly includes several million transistors per chip and is expected to reach 1 billion transistors. "Bio-chips" are being developed that will have the information processing capacity of human brain tissue. New CD-ROM technology will include read-write capability and significant increases in data capacity. Working prototypes of optical switches have been developed, and it is estimated that optical processors will operate up to 1,000 times faster than current microprocessors.

Events:
1. *Personal digital assistants (handheld microcomputers) are used by the majority of people to manage their work and personal affairs.*
Probability: 70 percent
Year: 2025
Leading nations: Japan, United States

2. *Supercomputers using massive parallel processing are commonly used. Thirty percent (of such computers) use this technology.*
Probability: 50 percent
Year: 2025
Leading nation: United States

3. PCs incorporate television, telephone, and interactive video transmission.

Probability: 40 percent

Year: 2000

Leading nation: United States

Comment: This is the new "information appliance" that virtually all computer hardware companies hope to produce in the near future. There is no doubt that it is technically feasible. The only question is whether an adequate market exists for it. If it does not reach the stores by 2000, it almost surely will do so within the following decade.

4. An entertainment center combining interactive television, telephone, and computing capability is commercially available for home use.

Probability: 75 percent

Year: 2025

Leading nations: United States, Japan

Comment: This is the next generation of the information appliance from scenario 3. Its practicality depends not only on the market, but on the ability of government and industry to set standards for what amounts to a new broadcast medium. If not for that consideration, it could arrive years earlier.

5. Optical computers enter the commercial marketplace.

Probability: 50 percent

Year: 2050

Leading nation: United States

Comment: We believe that the technical difficulty of making such a computer is greater than many forecasters recognize. Further, the growing power of more conventional technologies will slow the demand for optical computers, even in fields where the need for number-crunching power always outstrips the supply.

6. More advanced forms of data storage (optical, nonvolatile semiconductor, magnetic memory) are standard on multimedia PCs.
Probability: 50 percent
Year: 2025
Leading nation: United States

7. "Bio-chips" that store data in molecular bonds are commercially available.
Probability: 50 percent
Year: 2015
Leading nation: United States
Comment: If we generalize this to "some form of extremely dense memory system not yet available," the probability becomes much higher.

V. Information Technology; Computer Software

Confidence level: 60 percent

It is estimated that the use of expert systems is growing at a rate of 50 percent annually. Neural networks using parallel processors have achieved a technical breakthrough by permitting a superior form of computer architecture that solves problems more quickly, operates in a heuristic manner, recognizes patterns, and learns from its mistakes—much like the human brain. More powerful systems are being developed using thousands of multiple processors. Various forms of automated software development tools are being used widely, including computer-aided software engineering (CASE) tools and object-oriented programming. A variety of alternate entry systems, such as voice recognition, handwriting recognition, and optical scanning suggests a more user-friendly future.

Events:
1. The majority of software is generated automatically using software modules (object-oriented programming, CASE tools, and more).
Probability: 50 percent

Year: 2015

Leading nations: United States, Japan

2. Expert systems are routinely used to help in the decision-making process in management, medicine, engineering, and other fields.

Probability: 50 percent

Year: 2015

Leading nation: United States

3. Voice, handwriting, and optical recognition features allow ordinary PCs to interact with humans.

Probability: 50 percent.

Year: 2025

Leading nation: United States

Comment: Adequate single-user voice recognition is already available, of course. So are somewhat unreliable handwriting recognition systems. We interpret this question as referring to reliable systems that any user can deal with successfully, available as standard equipment, rather than as costly add-ons.

4. Computers are able to routinely translate languages in real-time with the accuracy and speed necessary for effective communications.

Probability: 50 percent

Year: 2010

Leading nation: United States

Comment: This equipment will reach general use when vocabularies open to untrained users reach six thousand words and three hundred idiomatic expressions.

5. Intelligent software agents (knowbots or navigators) routinely filter and retrieve information for users.

Probability: 50 percent

Year: 2025
Leading nation: United States

6. Ubiquitous computing environments (embedded processors in common objects) are integrated into the workplace and the home.
Probability: 50 percent
Year: 2015
Leading nation: United States

7. Thirty percent of computations are performed on neural networks using parallel processors.
Probability: 50 percent
Year: 2025
Leading nation: United States
Comment: However, we will not be surprised if these systems turn out to be less versatile than they now appear, and therefore account for less of the future computer market.

8. Computer programs are commonly available that can learn by trial and error in order to adjust their behavior (machine learning).
Probability: 50 percent
Year: 2015
Leading nation: United States
Comment: We interpret this question to refer to a high order of machine intelligence. Many programs could incorporate less sophisticated forms of "learning" by 2005, and conceivably sooner.

VI. Information Technology; Communications
Confidence level: 60 percent
Personal Communications Systems (PCS), the Internet, and the integration of all forms of common carriers (telephone, cable, and satellite) promise an interactive world that should connect every home and office.

The Internet is currently composed of 30 million users on more than 46,000 computer networks, and these figures will only increase. A low-cost network computer that downloads video content from the Internet is being developed.

Events:

1. PCS has a significant share of the market (10 percent) for voice communications.
Probability: 50 percent
Year: 2015
Leading nation: United States

2. Most communications systems (80 percent) in industrialized countries adopt a standard digital protocol.
Probability: 50 percent
Year: 2015
Leading nation: United States

3. Most people (80 percent) in developed countries access an "information superhighway."
Probability: 50 percent
Year: 2015
Leading nation: United States

4. Groupware systems are routinely used for simultaneously working and learning together at multiple sites.
Probability: 50 percent
Year: 2015
Leading nation: United States

5. Broadband networks (ISDN, ATM, and fiber optics) connect the majority of homes and offices.

Probability: 50 percent
Year: 2010
Leading nation: United States

VII. Information Technology; Information Services

Confidence level: 60 percent

Sophisticated information services are being developed that should offer more convenient and cost effective options for electronic education, working, shopping, and other functions that formerly required travel. For example, roughly 39 million people, or 35 percent of the U.S. work force work at home at least partially. Many retailers have created "Electronic Malls" that permit home shopping. It is estimated that sales of products through on-line services will soar from $200 million this year to $4.6 billion in 1998. Major universities are starting to use distance learning to create virtual universities.

Events:

1. A variety of movies, TV shows, sports, and other forms of entertainment can be selected electronically at home on demand.
Probability: 90 percent
Year: 2000
Leading nation: United States

2. Teleconferencing is routinely used in industrialized countries for business meetings.
Probability: 75 percent
Year: 2015
Leading nation: United States
Comment: The limiting factor here is not technology, but the purely human need for direct contact. Teleconferencing will be used as a kind of super-e-mail, in situations where people must cooperate or share in-

formation in real time, but do not need the reassurance and secondary cues of direct contact.

3. The majority of books and publications are published on-line.
Probability: 50 percent
Year: 2025
Leading nation: United States

4. Electronic banking, including electronic cash, replaces paper, checks, and cash as the principal means of commerce.
Probability: 50 percent
Year: 2025
Leading nation: United States

5. Half of all goods in the United States are sold through information services.
Probability: 25 percent
Year: 2025
Leading nation: United States

6. Most employees (80 percent) perform their jobs at least partially *from remote locations by telecommuting.*
Probability: 50 percent
Year: 2015
Leading nation: United States

7. Schools and colleges commonly use computerized teaching programs and interactive television lectures and seminars, as well as traditional methods.
Probability: 50 percent
Year: 2015
Leading nation: United States

VIII. Manufacturing and Robotics   Confidence level 90 percent

Automated factories (AF) incorporate robotics and computer systems to eliminate almost all human workers. The heart of the AF is computer-aided manufacturing (CIM), which unifies all manufacturing functions, including computer-aided design (CAD), computer-aided manufacturing (CAM), production scheduling, inventory control, operations, quality assurance, and so on into a single, fully automated system. The advantages of AF are considerable: large increases in productivity, almost perfect quality control, and the flexibility to customize small units of output. The obstacles are also considerable: displaced workers must be retrained to avoid labor-management conflict and unemployment, high capital investment costs are involved, and a different form of factory organization is required.

Events:

1. *Computer integrated manufacturing (CIM) is used in most (80 percent) factory operations.*
Probability: 50 percent
Year: 2015
Leading nations: United States, Japan

2. *Automation proceeds such that the proportion of factory jobs declines to less than 10 percent of the workforce.*
Probability: 50 percent
Year: 2015
Leading nation: United States

3. *Mass customization of products such as cars and appliances is commonly available [in 30 percent of products].*
Probability: 50 percent
Year: 2010
Leading nation: United States

4. Sophisticated robots that have sensory input, make decisions, learn, and are mobile become commercially available.

Probability: 50 percent

Year: 2050

5. Microscopic machines and/or nanotechnology are developed into commercial applications.

Probability: 50 percent

Year: 2050

Comment: This early in the field's development, we are uncomfortable making firm, specific forecasts about its chances of success. However, we are generally skeptical of any proposal that closely resembles true "atom-by-atom" nanotechnology. So far, nanotech's chief contribution has been that it inspires research in more conventional micro-scale technologies. That may turn out to be its permanent role.

IX. Materials   Confidence level: 80 percent

Materials are being developed to serve almost any possible purpose. Plastic composites can be made as tough as steel, to conduct electricity and light, and to be biodegradable. Ceramics and alloys of metals offer high strength, low weight, temperature resistance, and superconductivity. Intelligent materials are being developed that adjust their properties to environmental changes and are self-repairing, including actuators, sensors, and control microprocessors. Self-assembling materials, which do not require human intervention, are a concept inspired by nature and generating increasing interest. The significance of these advances is that there may soon exist the ability to design almost any type of customized material to suit the product designer's specifications.

Events:

1. Ceramic engines are mass produced for commercial vehicles.

Probability: 50 percent

Year: 2050
Leading nation: United States

2. Half of all automobiles are made of recyclable plastic components.
Probability: 50 percent
Year: 2025
Leading nation: United States

3. Superconducting materials are commonly used (in 30 percent of equipment that potentially might employ them) for transmitting electricity in electronic devices, such as energy, medical, and communications applications.
Probability: 50 percent
Year: 2025
Leading nations: United States, Japan

4. Material compositions replace the majority of traditional metals in product design.
Probability: 50 percent
Year: 2025
Leading nation: United States

5. "Buckyballs" or "buckytubes" are instrumental in developing new materials.
Comment: Though these materials are fascinating and clearly appear to hold promise, we believe it is too early to make this kind of forecast. We just do not know enough about them to be certain of their future applications.

6. Self-assembling materials are routinely used commercially.
Probability: 50 percent

Year: 2050
Leading nation: United States

7. Intelligent materials are routinely used in homes, offices, and vehicles.
Probability: 50 percent
Year: 2050
Leading nation: United States

X. Medicine   Confidence level: 75 percent

The present mapping of the entire human genome and the ability to manipulate genetic codes should make it possible to cure hereditary diseases, produce genetically improved plants and animals, and create more powerful drugs and vaccines. Gene therapy now allows physicians to treat many diseases by injecting genes directly into the body. In 1994, more than 200 people were treated with therapeutic genes in twelve trials. The growing use of transplanted and artificial organs is rapidly reaching the point where almost all parts of the human body can be replaced, including portions of the brain and central nervous system. Scientists are now growing skin, bone, and cartilage for surgical replacement procedures. Computerized information systems, including the use of expert systems, suggest that routine medical functions may be automated to greatly facilitate research, improve treatment, and hold down costs. Thirty telemedicine networks exist already in the United States. Evidence is mounting that most illnesses today are the result of diet, stress, lack of exercise, environmental pollutants, lifestyle, attitude, and other "soft" factors; in response, holistic health care seems to be gaining recognition by the medical profession.

Events:
1. Computerized information systems are commonly used for medical care, including diagnosis, dispensing prescriptions, monitoring medical conditions, and promoting self-care.

Probability: 75 percent

Year: 2005

Leading nation: United States

2 Holistic (physical and mental) approaches to health care become accepted by the majority of the medical community.

Probability: 50 percent

Year: 2000

Leading nation: United States

3. Parents can routinely choose characteristics of their children through genetic engineering.

Probability: 50 percent

Year: 2025

Leading nation: United States

Comment: Prenatal genetic intervention will be technically feasible, and it seems clear that parents will choose to have their children healed of genetic disorders before birth the moment the technology becomes available. However, many practices that are technically feasible will remain extremely controversial. These include choosing the child's sex and "fine-tuning" such factors as eye and hair color. It is likely that these matters will be left to chance well beyond the next thirty years.

4. Gene therapy is routinely used to prevent and/or cure an inherited disease.

Probability: 50 percent

Year: 2015

Leading nation: United States

5. Living organs and tissue produced genetically are routinely used for replacement.

Probability: 50 percent

Year: 2010

Leading nation: United States

6. Artificial organs and tissue produced synthetically are routinely used for replacement.

Probability: 50 percent

Year: 2025

Leading nation: United States

7. Computerized vision is commercially available to correct eye defects.

Probability: 50 percent

Year: 2005

Leading nation: United States

Comment: However, the development of functional eye transplants could quickly supplant this technology.

8. A cure or preventive for a major disease such as cancer or AIDS is found.

Probability: 50 percent

Year: 2015

Leading nation: United States, France

Comment: Any one "cure" for AIDS is likely to be effective only against certain strains of the disease.

(Authors' note: As described in Chapter 9, it seems that something approaching a "cure for cancer" may already have been discovered. Recent reports of success with multidrug AIDS treatments suggest that we may have been too pessimistic about the possibility of a generally effective treatment for this infection.)

XI. Space   Confidence level: 50 percent

Several new space technologies may emerge for wide use in the near future. Orbiting laboratories are in various stages of planning that would conduct scientific experiments and produce exotic industrial materials and products. Alternatives to rocket launchers are being developed that would fire payloads into orbit using an electromagnetic "rail gun." Private entrepreneurs are entering the industry to launch vehicles and conduct other aspects of space research and development. New propulsion technologies will increase both the speed and distance of space voyages. Programs are under way to develop closed ecological systems that would sustain the lives of astronauts for long voyages.

Events:

1. Private corporations perform the majority of space launches as private ventures.

Probability: 50 percent

Year: 2025

Leading nation: United States

2. A manned mission to Mars is completed.

Probability: 50 percent

Year: 2050

Leading nation: United States

Comment: There is no technical reason this could not be completed much before we project. However, not even the discovery of evidence that life once existed on Mars has given political and scientific leaders the will to undertake such an ambitious mission. In fact, any trace of hope that life might exist on another planet offers a powerful argument against visiting it in person. Robotic landers can be sterilized, while a manned mission would risk infecting an alien ecosystem with earthly bacteria. This conceivably could destroy native species and surely would complicate the task of studying them. Some astonishing discovery by Mars probes in the next ten to fifteen years could renew interest in a manned mission. Until it actually occurs, hope for a manned flight to Mars will remain dim.

3. A permanently manned moon base is established.
Probability: 50 percent
Year: 2025
Leading nation: United States
Comment: Again, this is technically possible, and it would be much cheaper than any conceivable Mars landing. However, the will to undertake any such ambitious mission is absent, and there is no excuse to hope that it will soon develop.

4. A spaceship is launched to explore a neighboring star system.
Probability: 50 percent
Year: 2025
Leading nations: United States, Russia

5. Chemicals and metals that are not feasible on Earth due to purity and other requirements are developed in space.
Probability: 50 percent
Year: 2010
Leading nation: United States

6. Spaceships or probes approach (80 percent) the speed of light.
Comment: Though on a theoretical level this is a surprisingly modest goal, even at 1 G acceleration, it lies far beyond anything that has yet been seriously proposed. We believe it is too early to make any forecast about this scenario.

7. Intelligent life is contacted elsewhere in the universe.
Comment: Again, no serious forecast is possible.

XII. Transportation   Confidence level: 70 percent
    Several advanced modes of transportation may alleviate the estimated $100 billion a year that Americans waste due to traffic congestion. Mag-

netic levitation (maglev) trains, running on a cushion of air at speeds of up to 300 mph, are being developed in Japan and several American states. (Authors' note: Here in the United States, maglev development projects now are on hold.) When operational, trips between major cities could be made in about one hour from city centers. A variety of automated highway systems are being developed that control steering, braking, and navigation. General Motors is caravaning cars to travel at speeds of up to 70 mph at distances of six inches from bumper to bumper. Computer systems are now available to guide traffic along the most favorable route, cutting commuting time. The Intelligent Transportation Society estimates that $209 billion will be spent between now and 2011 on intelligent highway systems. The Aerospace Plane is being developed to permit hypersonic air travel.

Events:

1. *High-speed rail or maglev trains are available between most major cities in developed countries.*
Probability: 50 percent
Year: 2050
Leading nations: United States, France
Comment: As is often the case, this development will be slowed by economic and political concerns, not by technological hurdles.

2. *Hybrid vehicles (electric and internal combustion engine) are commercially available.*
Probability: 50 percent
Year: 2025
Leading nation: United States
Comment: We interpret this question to mean "commercially available on a scale large enough to have a significant impact on the automotive market." They could be available from tiny start-up companies or as costly options from the major automobile manufacturers by 2005 or so, but will have only a negligible impact on the sale of conventional cars.

3. Battery-powered electric cars are commonly available accounting for 30 percent of the auto market.
Probability: 50 percent
Year: 2050
Leading nation: United States

4. Fuel cell-powered electric cars are commonly available (accounting for 30 percent of the auto market).
Probability: 50 percent
Year: 2075
Leading nation: United States

5. Hypersonic planes are used for the majority of transoceanic flights.
Probability: 50 percent
Year: 2050
Leading nation: United States

6. Automated highway systems are commonly used to reduce highway congestion (on 30 percent of major highways in congested areas).
Probability: 50 percent
Year: 2025
Leading nation: United States

7. Intelligent transportation systems are commonly used to reduce highway congestion (in 30 percent of suitable areas).
Probability: 50 percent
Year: 2015
Leading nations: United States, Japan

8. Personal rapid transit (such as car-like capsules on guided rails) are installed in most metropolitan areas.
Probability: 50 percent

Year: 2050
Leading nation: United States

9. Clustered, self-contained communities in urban areas reduce the need for local transportation.
Probability: 25 percent
Year: 2050
Leading nation: United States

TABLE 2

A CONSENSUS VIEW OF THE FUTURE

I. Energy

Alternative sources for fossil fuels are now becoming feasible, as the cost of conventional energy and its resulting environmental damage grows. Photovoltaic cells are now commonly used for special purposes, such as calculators, billboard lights, battery chargers, and experimental solar-powered automobiles, and they are starting to be used for homes and power plants. Luz International Company operates a state-of-the-art solar power plant in the Mojave desert that produces electricity at comparable costs to conventional sources. Thousands of windmills are now producing energy in the United States. More efficient lights are available that cut energy use by more than 50 percent. More efficient fuel cells using gasoline, natural gas, and hydrogen are being developed for commercial use. According to the U.S. Department of Energy, renewable energy supplies about 10 percent of United States energy, and this figure should increase to 28 percent by 2030. Fission nuclear power is undergoing an important transition, as reactor designs become fail-safe and small enough to power large buildings.

Events:

1. A significant portion (10 percent) of energy usage is derived from alternative energy sources, such as geothermal, hydroelectric, solar/photovoltaic.
Probability: 77 percent
Year: 2010
Demand: $45.6 billion
Leading nation: United States

2. Energy efficiency improves by 50 percent through innovations in transportation, industrial processing, and environmental control.
Probability: 61 percent
Year: 2016
Demand: $49.2 billion
Leading nation: United States

3. Fuel cells, converting fuels to electricity, are commonly used. ([They provide] 30 percent [of the electric power generated.])
Probability: 53 percent
Year: 2017
Demand: $61.2 billion
Leading nation: United States

4. Biological materials, such as crops, trees, and other forms of organic matter, are used as significant (10 percent) energy sources.
Probability: 60 percent
Year: 2011
Demand: $43 billion
Leading nation: United States

5. Fission nuclear power is used for at least 50 percent of electricity generation.

Probability: 46 percent
Year: 2020
Demand: $26.3 billion
Leading nation: France

6. *Fusion nuclear power is used commercially for electricity production.*
Probability: 50 percent
Year: 2026
Demand: $113.3 billion
Leading nation: United States

7. *Hydrogen becomes routinely used in energy systems.*
Probability: 50 percent
Year: 2020
Demand: $102 billion
Leading nation: United States

II. Environment

Global warming, acid rain, ozone depletion, and other such threats have caused government officials, business leaders, and households to rate the environment as the biggest issue facing this decade. Many corporations are voluntarily recovering manufacturing wastes to avoid pollution and to recapture valuable byproducts, and they are redesigning products and packaging to reduce environmental impacts. For instance, Dow Corning saved more than $500 million through its "Pollution Prevention Pays Program," and DuPont is planning to curb all environmental emissions from its products. Japan recycles 50 percent of all household wastes, and the concept is being adopted in most American cities. As a result, the proportion of steel, aluminum, glass, paper, and other products composed of recycled materials is growing. Industrial ecology focuses on recycling and reusing byproducts within and between industries.

Events:

1. The majority of CFCs are replaced by materials that do not damage the ozone layer, such as HFCs.

Probability: 77 percent

Year: 2006

Demand: $52.2 billion

Leading nation: United States

2. One-half of the waste from households in developed countries is recycled.

Probability: 74 percent

Year: 2008

Demand: $53.2 billion

Leading nations: United States, Japan

3. Most manufacturers adopt "green" methods that minimize environmental pollution.

Probability: 73 percent

Year: 2010

Demand: $90 billion

Leading nation: United States

4. The majority of manufactured goods use recycled materials.

Probability: 66 percent

Year: 2016

Demand: $126 billion

Leading nation: Germany

5. Improvements in fossil fuel energy efficiency and greater use of alternative energy sources reduce "greenhouse" gas emission by one-half from current volumes.

Probability: 59 percent

Year: 2016
Demand: $46 billion
Leading nation: United States

6. *The majority of manufacturing facilities use industrial ecology (eco-industrial parks operating as a closed system) to reduce waste and pollution.*
Probability: 55 percent
Year: 2015
Demand: $48 billion
Leading nation: United States

III. Farming and Food

Hydroponic farming is increasing crop yields and can be used anywhere. Food is being manufactured from raw materials to produce artificial foods, such as the use of cellulose from wood to make bread and meats. Aquaculture now provides about 22 percent of fish consumption, and is showing rapid growth as the demand for seafood continues to rise because of trends in health consciousness and medical research. The enormous potential for aquaculture is suggested by the fact that in 1991 livestock production was a $70 billion market, while aquaculture only accounted for $1 billion. In 1995, aquaculture is a $26 billion industry. New food derivations from sea vegetation are being developed. Advances in biotechnology have produced the first commercially available, genetically altered vegetables that grow faster, resist disease, and are more robust. Alternative/organic farming methods, used to rejuvenate soil, control crop pests, and recycle organic materials, are being adopted by many farms, especially as the use of fertilizers and traditional farming methods becomes increasingly expensive and less effective.

Events:

1. *Genetic engineering techniques are routinely used to produce new strains of plants and animals.*

Probability: 75 percent
Year: 2008
Demand: $66.8 billion
Leading nation: United States

2. The use of chemical fertilizers and pesticides declines by half.
Probability: 60 percent
Year: 2012
Demand: $27.5 billion
Leading nation: United States

3. The majority of farming in industrialized countries incorporates alternative/organic farming techniques into traditional methods.
Probability: 57 percent
Year: 2015
Demand: $76 billion
Leading nation: United States

4. Seafood grown using aquaculture provides the majority of seafood consumed.
Probability: 56 percent
Year: 2014
Demand: $52.4 billion
Leading nation: Japan

5. Automation of farming methods, using technology such as robotics, is common. Thirty percent [of farming employs some such technology].
Probability: 60 percent
Year: 2020
Demand: $82.4 billion
Leading nation: United States

6. Precision farming (computerized control of irrigation, seeding, fertilizer, and pesticides) is common. Thirty percent [of farming employs some such technology].
Probability: 69 percent
Year: 2015
Demand: $71.4 billion
Leading nation: United States

7. Urban production of fruits and vegetables using greenhouses and/or other intensive production systems is common. Thirty percent or more [of fruits and vegetables are grown in such "farms"].
Probability: 53 percent
Year: 2020
Demand: $55 billion
Leading nation: United States

8. Produce grown using hydroponic methods is common. (It accounts for 30 percent or more of all produce.)
Probability: 53 percent
Year: 2015
Demand: $40 billion
Leading nation: United States

9. Artificial meats, vegetables, bread, etc., are commonly consumed. (They account for 30 percent of all such products.)
Probability: 39 percent
Year: 2022
Demand: $74.8 billion
Leading nation: United States

IV. Information Technology; Computer Hardware
There seems to be no end in sight to the forty-year trend in which computer power has increased about 50 percent per year, or by a factor

of ten with each successive generation of machines. Microcircuitry now commonly includes several million transistors per chip and is expected to reach 1 billion transistors. "Bio-chips" are being developed that will have the information processing capacity of human brain tissue. New CD-ROM technology will include read-write capability and significant increases in data capacity. Working prototypes of optical switches have been developed, and it is estimated that optical processors will operate up to 1,000 times faster than current microprocessors.

Events:
1. Personal digital assistants (handheld microcomputers) are used by the majority of people to manage their work and personal affairs.
Probability: 75 percent
Year: 2008
Demand: $53.8 billion
Leading nation: United States

2. Supercomputers using massive parallel processing are commonly used. Thirty percent [of such computers use this technology].
Probability: 80 percent
Year: 2008
Demand: $63.6 billion
Leading nation: United States

3. PCs incorporate television, telephone, and interactive video transmission.
Probability: 84 percent
Year: 2005
Demand: $111.4 billion
Leading nation: United States

4. An entertainment center combining interactive television, telephone, and computing capability is commercially available for home use.

Probability: 83 percent
Year: 2006
Demand: $108.6 billion
Leading nation: United States

5. Optical computers enter the commercial marketplace.
Probability: 64 percent
Year: 2014
Demand: $67.1 billion
Leading nation: United States

6. More advanced forms of data storage (optical, nonvolatile semiconductor, and magnetic memory) are standard on multimedia PCs.
Probability: 75 percent
Year: 2006
Demand: $43.6 billion
Leading nation: United States

7. "Bio-chips" that store data in molecular bonds are commercially available.
Probability: 54 percent
Year: 2017
Demand: $57.8 billion
Leading nation: United States

V. Information Technology; Computer Software

It is estimated that the use of expert systems is growing at a rate of 50 percent annually. Neural networks using parallel processors have achieved a technical breakthrough by permitting a superior form of computer architecture that solves problems more quickly, operates in a heuristic manner, recognizes patterns, and learns from its mistakes—much like the human brain. More powerful systems are being developed using thousands of multiple processors. Various forms of automated software development tools are being used widely, including computer-aided software engineering (CASE) tools and object-oriented programming. A variety of alternate entry systems, such as voice recognition, handwriting recognition, and optical scanning suggests a more user-friendly future.

Events:

1. The majority of software is generated automatically using software modules (object-oriented programming, CASE tools).
Probability: 72 percent
Year: 2007
Demand: $47.1 billion
Leading nations: United States, Japan

2. Expert systems are routinely used to help in the decision-making process in management, medicine, engineering, and other fields.
Probability: 72 percent
Year: 2010
Demand: $59.3 billion
Leading nation: United States

3. Voice, handwriting, and optical recognition features allow ordinary PCs to interact with humans.
Probability: 73 percent.
Year: 2007
Demand: $34.3 billion
Leading nation: United States

4. Computers are able to routinely translate languages in real-time with the accuracy and speed necessary for effective communications.
Probability: 65 percent
Year: 2012

Demand: $40.9 billion
Leading nation: United States

5. Intelligent software agents (knowbots or navigators) routinely filter and retrieve information for users.
Probability: 79 percent
Year: 2009
Demand: $28.5 billion
Leading nation: United States

6. Ubiquitous computing environments (embedded processors in common objects) are integrated into the workplace and the home.
Probability: 75 percent
Year: 2009
Demand: $32.5 billion
Leading nation: United States

7. Thirty percent of computations are performed on neural networks using parallel processors.
Probability: 61 percent
Year: 2015
Demand: $28.5 billion
Leading nation: United States

8. Computer programs are commonly available that can learn by trial and error in order to adjust their behavior (machine learning).
Probability: 67 percent
Year: 2012
Demand: $31 billion
Leading nation: United States

VI. Information Technology; Communications

Personal Communications Systems (PCS), the Internet, and the integration of all forms of common carriers (telephone, cable, and satellite) promise an interactive world that should connect every home and office. The Internet is currently composed of 30 million users on more than 46,000 computer networks, and these figures will only increase. A low-cost network computer that downloads video content from the Internet is being developed.

Events:

1. PCS has a significant share of the market (10 percent) for voice communications.
Probability: 56 percent
Year: 2006
Demand: $42.1 billion
Leading nation: United States

2. Most communications systems (80 percent) in industrialized countries adopt a standard digital protocol.
Probability: 70 percent
Year: 2006
Demand: $70 billion
Leading nation: United States

3. Most people (80 percent) in developed countries access an "information superhighway."
Probability: 78 percent
Year: 2008
Demand: 74.3 billion
Leading nation: United States

4. Groupware systems are routinely used for simultaneously working and learning together at multiple sites.
Probability: 75 percent
Year: 2007
Demand: $33.3 billion
Leading nation: United States

5. Broadband networks (ISDN, ATM, fiber optics) connect the majority of homes and offices.
Probability: 70 percent
Year: 2009
Demand: $103.3 billion
Leading nation: United States

VII. Information Technology; Information Services

Sophisticated information services are being developed that should offer more convenient and cost-effective options for electronic education, working, shopping, and other functions that formerly required travel. For example, roughly 39 million people, or 35 percent of the U.S. work force work at home at least partially. Many retailers have created "Electronic Malls" that permit home shopping. It is estimated that sales of products through on-line services will soar from $200 million this year to $4.6 billion in 1998. Major universities are starting to use distance learning to create virtual universities.

Events:
1. A variety of movies, TV shows, sports, and other forms of entertainment can be selected electronically at home on demand.
Probability: 84 percent
Year: 2003
Demand: $90 billion
Leading nation: United States

2. Teleconferencing is routinely used in industrialized countries for business meetings.
Probability: 83 percent
Year: 2004
Demand: $44.5 billion
Leading nation: United States

3. The majority of books and publications are published on-line.
Probability: 60 percent
Year: 2013
Demand: $65.8 billion
Leading nation: United States

4. Electronic banking, including electronic cash, replaces paper, checks, and cash as the principal means of commerce.
Probability: 70 percent
Year: 2009
Demand: $69 billion
Leading nation: United States

5. Half of all goods in the United States are sold through information services.
Probability: 55 percent
Year: 2018
Demand: $208.3 billion
Leading nation: United States

6. Most employees (80 percent) perform their jobs at least partially *from remote locations by telecommuting.*
Probability: 56 percent
Year: 2014
Demand: $468 billion
Leading nation: United States

7. Schools and colleges commonly use computerized teaching programs and interactive television lectures and seminars, as well as traditional methods.
Probability: 78 percent
Year: 2006
Demand: $41.5 billion
Leading nation: United States

VIII. Manufacturing and Robotics

Automated factories (AF) incorporate robotics and computer systems to eliminate almost all human workers. The heart of the AF is computer-aided manufacturing (CIM), which unifies all manufacturing functions, including computer-aided design (CAD), computer-aided manufacturing (CAM), production scheduling, inventory control, operations, and quality assurance, into a single, fully automated system. The advantages of AF are considerable: large increases in productivity, almost perfect quality control, and the flexibility to customize small units of output. The obstacles are also considerable: displaced workers must be retrained to avoid labor-management conflict and unemployment, high capital investment costs are involved, and a different form of factory organization is required.

Events:
1. Computer integrated manufacturing (CIM) is used in most (80 percent) factory operations.
Probability: 73 percent
Year: 2012
Demand: $124 billion
Leading nations: United States, Japan

2. Automation proceeds such that the proportion of factory jobs declines to less than 10 percent of the workforce.
Probability: 67 percent

Year: 2015
Demand: $150 billion
Leading nation: United States

3. Mass customization of products such as cars and appliances is commonly available (in 30 percent of products).
Probability: 73 percent
Year: 2011
Demand: $330 billion
Leading nation: United States

4. Sophisticated robots that have sensory input, make decisions, learn, and are mobile become commercially available.
Probability: 64 percent
Year: 2016
Demand: $130 billion
Leading nation: Japan

5. Microscopic machines and/or nanotechnology are developed into commercial applications.
Probability: 66 percent
Year: 2016
Demand: $31.3 billion
Leading nation: United States

IX. Materials

Materials are being developed to serve almost any possible purpose. Plastic composites can be made as tough as steel, to conduct electricity and light, and to be biodegradable. Ceramics and alloys of metals offer high strength, low weight, temperature resistance, and superconductivity. Intelligent materials are being developed that adjust their properties to environmental changes and are self-repairing, including actuators, sensors, and control microprocessors. Self-assembling materials, which do not require human intervention, are a concept inspired by nature and generating increasing interest. The significance of these advances is that there may soon exist the ability to design almost any type of customized material to suit the product designer's specifications.

Events:

1. *Ceramic engines are mass produced for commercial vehicles.*
Probability: 58 percent
Year: 2014
Demand: $49 billion
Leading nation: Japan

2. *Half of all automobiles are made of recyclable plastic components.*
Probability: 58 percent
Year: 2013
Demand: $51 billion
Leading nation: United States

3. *Superconducting materials are commonly used (in 30 percent of equipment that potentially might employ them) for transmitting electricity in electronic devices, such as energy, medical, and communications applications.*
Probability: 56 percent
Year: 2015
Demand: $43 billion
Leading nation: United States

4. *Material composites replace the majority of traditional metals in product design.*
Probability: 53 percent
Year: 2016

Demand: $100 billion
Leading nation: United States

5. "Buckyballs" or "buckytubes" are instrumental in developing new materials.
Probability: 59 percent
Year: 2011
Demand: $20 billion
Leading nation: United States

6. Self-assembling materials are routinely used commercially.
Probability: 56 percent
Year: 2027
Demand: $82 billion
Leading nation: United States

7. Intelligent materials are routinely used in homes, offices, and vehicles.
Probability: 57 percent
Year: 2026
Demand: $66 billion
Leading nation: United States

X. Medicine

The present mapping of the entire human genome and the ability to manipulate genetic codes should make it possible to cure hereditary diseases, produce genetically improved plants and animals, and create more powerful drugs and vaccines. Gene therapy now allows physicians to treat many diseases by injecting genes directly into the body. In 1994, more than two hundred people were treated with therapeutic genes in twelve trials. The growing use of transplanted and artificial organs is rapidly reaching the point where almost all parts of the human body can be replaced, including portions of the brain and central nervous system.

Scientists are now growing skin, bone, and cartilage for surgical replacement procedures. Computerized information systems, including the use of expert systems, suggest that routine medical functions may be automated to greatly facilitate research, improve treatment, and hold down costs. Thirty telemedicine networks exist already in the United States. Evidence is mounting that most illnesses today are the result of diet, stress, lack of exercise, environmental pollutants, lifestyle, attitude, and other "soft" factors; in response, holistic health care seems to be gaining recognition by the medical profession.

Events:

1. Computerized information systems are commonly used for medical care, including diagnosis, dispensing prescriptions, monitoring medical conditions, and promoting self-care.
Probability: 82 percent
Year: 2007
Demand: $87 billion
Leading nation: United States

2. Holistic (physical and mental) approaches to health care become accepted by the majority of the medical community.
Probability: 61 percent
Year: 2009
Demand: $55 billion
Leading nation: United States

3. Parents can routinely choose characteristics of their children through genetic engineering.
Probability: 53 percent
Year: 2020
Demand: $21.3 billion
Leading nation: United States

4. Gene therapy is routinely used to prevent and/or cure an inherited disease.
Probability: 63 percent
Year: 2013
Demand: $63.3 billion
Leading nation: United States

5. Living organs and tissue produced genetically are routinely used for replacement.
Probability: 53 percent
Year: 2018
Demand: $62.7 billion
Leading nation: United States

6. Artificial organs and tissue produced synthetically are routinely used for replacement.
Probability: 58 percent
Year: 2019
Demand: $68.3 billion
Leading nation: United States

7. Computerized vision is commercially available to correct eye defects.
Probability: 56 percent
Year: 2014
Demand: $31.7 billion
Leading nation: United States

8. A cure or preventive for a major disease such as cancer or AIDS is found.
Probability: 58 percent
Year: 2013
Demand: $116 billion
Leading nation: United States

XI. Space

Several new space technologies may emerge for wide use in the near future. Orbiting laboratories are in various stages of planning that would conduct scientific experiments and produce exotic industrial materials and products. Alternatives to rocket launchers are being developed that would fire payloads into orbit using an electromagnetic "rail gun." Private entrepreneurs are entering the industry to launch vehicles and conduct other aspects of space research and development. New propulsion technologies will increase both the speed and distance of space voyages. Programs are under way to develop closed ecological systems that would sustain the lives of astronauts for long voyages.

Events:
1. *Private corporations perform the majority of space launches as private ventures.*
Probability: 62 percent
Year: 2013
Demand: $60 billion
Leading nation: United States

2. *A manned mission to Mars is completed.*
Probability: 59 percent
Year: 2029
Demand: $30 billion
Leading nation: United States

3. *A permanently manned moon base is established.*
Probability: 55 percent
Year: 2028
Demand: $32.5 billion
Leading nation: United States

4. *A spaceship is launched to explore a neighboring star system.*
Probability: 51 percent

Year: 2042
Demand: $47.5 billion
Leading nation: United States

5. Chemicals and metals that are not feasible on Earth due to purity and other requirements are developed in space.
Probability: 57 percent
Year: 2012
Demand: $21.5 billion
Leading nation: United States

6. Spaceships or probes approach (80 percent) the speed of light.
Probability: 43 percent
Year: 2062
Demand: $75.5 billion
Leading nation: United States

7. Intelligent life is contacted elsewhere in the universe.
Probability: 33 percent
Year: 2049
Demand: $45.3 billion
Leading nation: United States

XII. Transportation

Several advanced modes of transportation may alleviate the estimated $100 billion a year that Americans waste due to traffic congestion. Magnetic levitation (maglev) trains, running on a cushion of air at speeds of up to 300 mph, are being developed in Japan and several American states. (Authors' note: Here in the United States, maglev development projects now are on hold.) When operational, trips between major cities could be made in about one hour from city centers. A variety of automated highway systems are being developed that control steering, braking, and navigation. General Motors is caravaning cars to travel at speeds of up to 70 mph at distances of six inches from bumper to bumper. Computer systems are now available to guide traffic along the most favorable route, cutting commuting time. The Intelligent Transportation Society estimates that $209 billion will be spent between now and 2011 on intelligent highway systems. The Aerospace Plane is being developed to permit hypersonic air travel.

Events:

1. *High-speed rail or maglev trains are available between most major cities in developed countries.*
Probability: 58 percent
Year: 2017
Demand: $120 billion
Leading nation: Japan

2. *Hybrid vehicles (electric and internal combustion engine) are commercially available.*
Probability: 69 percent
Year: 2006
Demand: $87 billion
Leading nation: United States

3. *Battery-powered electric cars are commonly available (accounting for 30 percent of the auto market).*
Probability: 70 percent
Year: 2011
Demand: $102 billion
Leading nation: United States

4. *Fuel cell-powered electric cars are commonly available (accounting for 30 percent of the auto market).*
Probability: 58 percent

Year: 2016
Demand: $116 billion
Leading nation: United States

5. Hypersonic planes are used for the majority of transoceanic flights.
Probability: 48 percent
Year: 2025
Demand: $91 billion
Leading nation: United States

6. Automated highway systems are commonly used to reduce highway congestions [on 30 percent of major highways in congested areas].
Probability: 55 percent
Year: 2018
Demand: $70 billion
Leading nation: United States

7. Intelligent transportation systems are commonly used to reduce highway congestion [in 30 percent of suitable areas].
Probability: 58 percent
Year: 2016
Demand: $90 billion
Leading nation: United States

8. Personal rapid transit (such as car-like capsules on guided rails) are installed in most metropolitan areas.
Probability: 43 percent
Year: 2024
Demand: $62.5 billion
Leading nation: United States

9. Clustered, self-contained communities in urban areas reduce the need for local transportation.
Probability: 53 percent
Year: 2023
Demand: $85 billion
Leading nation: United States

TABLE 3

A TIMELINE FOR THE EVOLUTION OF

TECHNOLOGY

2003 — Entertainment on Demand
2004 — Videoconferencing
2005 — PC Convergence
2006 — Advanced Data Storage
        CFCs Are Replaced
        Distance Learning
        Entertainment Centers
        Hybrid Vehicles Common
        PCS Gains Markets
        Standard Digital Protocol
2007 — Computer Sensory Recognition
        Computerized Self-Care
        Groupware Systems
        Modular Software
2008 — Genetically Produced Food
        Half of Household Waste Recycled
        Information Superhighway
        Parallel Processing Computing
        Personal Digital Assistants
2009 — Broadband Networks
        Electronic Banking/Cash
        Holistic Health Care
        Intelligent Agents
        Ubiquitous Computing Environment
2010 — Alternative Energy Sources
        Expert Systems
        "Green" Environmental Methods

2011 — Buckyballs and Buckytubes
      Electric Cars Are Common
      Mass Customization
      Organic Energy Sources
2012 — CIM Used in Most Factories
      Computer Language Translation
      Farm Chemicals Drop by Half
      Machine Learning
2013 — Gene Therapy
      Half of Autos Recyclable
      Major Diseases Cured
      On-Line Publishing
2014 — Aquaculture
      Ceramic Engines
      Computerized Vision Implants
      Optical Computers
2015 — Alternative/Organic Farming
      Factory Jobs Drop to 10 Percent
      Hydroponic Produce
      Industrial Ecology
      Neural Networks
      Precision Farming
      Superconducting Materials
2016 — Energy Efficiency
      Fossil Fuels Cut Greenhouse Gas
      Fuel Cell Electric Cars
      Intelligent Transportation Systems
      Material Composites
      Nanotechnology
      Recycled Goods
      Sophisticated Robots

2017 — Biochips
       Fuel Cells
       High-Speed Maglev
2018 — Automated Highway Systems
       Cloning/Organ Replacement
       Half of Goods Sold Electronically
       New Materials from Space
2019 — Private Space Ventures
       Synthetic Body Parts
       Telecommuting
2020 — Farm Automation
       Fission Power
       Genetic Engineering
       Hydrogen Energy
       Urban Greenhouses
2022 — Artificial Foods
2023 — Clustered Communities
2024 — Personal Rapid Transit
2025 — Hypersonic Air Travel
2026 — Fusion Power
       Intelligent Materials
2027 — Self-Assembling Materials
2028 — Permanent Moon Base
2037 — Manned Mars Mission
2042 — Stellar Exploration
2049 — Extraterrestrial Contact
2062 — Near-Light Speed Achieved

TABLE 4

CONTRIBUTORS TO THE 1996 DELPHI SURVEY

OF EMERGING TECHNOLOGIES

RESPONDENT	SPECIALTY	AFFILIATION
Claudio D. Antonini	Nuclear Engineer	N/A
Thomas Arrison	Research Assistant	Office of Japan Affairs, National Research Council
John Artz	Assistant Professor of Management Science	George Washington University
Mohsen Bahrami	Mechanical Engineering Department	Amir Kaabir University of Technology
Marvin J. Cetron	President	Forecasting International
Travis P. Charbeneau	N/A	N/A
Tom Erickson	Management Data Systems	Lockheed Martin
Ted Gordon	Futurist	N/A
William E. Halal	Professor of Management	George Washington University
Olaf Helmer	N/A	N/A
Maj. George Hluck	Artificial Intelligence	N/A
David Jensen	Professor of Computer Science	University of Massachusetts—Amherst
Michael D. Kull	Doctoral Candidate	Management Science Dept., George Washington University
Ann Leffmann	Doctoral Candidate	Management Science Dept., George Washington University
Harold A. Linstone	Systems Science Ph.D. Program	Portland State University
Spyros G. Makridakis	Research Professor	INSEAD
Sarfraz A. Mian	Assistant Professor	School of Business, SUNY—Oswego
Tomonobu Noguchi	Electronic and Information	Dow Corning Toray Silicone Industries Co., Ltd.
John Petersen	Futurist	N/A
Peter Rzeszotarski	Environmental Science	Military
Paul J. Werbos	Research Program Director	National Science Foundation
Eberhard Weber	Research Project on Global Systems	N/A

INDEX